African Philosophy and Environmental Conservation

African Philosophy and Environmental Conservation is about the unconcern for, and marginalisation of, the environment in African philosophy. The issue of the environment is still very much neglected by governments, corporate bodies, academics and, specifically, philosophers in sub-Saharan Africa. The entrenched traditional worldviews which give a place of privilege to one thing over the other, as, for example, men over women, is the same attitude that privileges humans over the environment. This culturally embedded orientation makes it difficult for stakeholders in Africa to identify and confront the modern-day challenges posed by the neglect of the environment. In a continent where deep-rooted cultural and religious practices, as well as widespread ignorance, determine human conduct towards the environment, it becomes difficult to curtail, much less overcome, the threats to our environment. It shows that, to a large extent, the African cultural privileging of men over women and of humans over the environment somewhat exacerbates and makes the environmental crisis on the continent intractable. For example, it raises the challenging puzzle as to why women in Africa are the ones to plant the trees and men are the ones to fell them.

Contributors address these salient issues from both theoretical and practical perspectives, demonstrating what African philosophy could do to ameliorate the marginalisation which the theme of environment suffers on the continent. Philosophy is supposed to teach us how to lead the good life in all its forms; why is it failing in this duty in Africa specifically insofar as the issue of the environment is concerned?

This book, which trail-blazes the field of African Philosophy and Environmental Ethics, will be of great interest to students and scholars of Philosophy, African philosophy, Environmental Ethics and Gender Studies.

Jonathan Okeke-Mpi Chimakonam is a senior lecturer at the Department of Philosophy, University of Calabar, Nigeria, and Research Fellow, Department of Philosophy, University of Pretoria, South Africa.

Routledge Explorations in Environmental Studies

www.routledge.com/Routledge-Explorations-in-Environmental-Studies/book-series/REES

Socioeconomic Evaluation of Megaprojects
Dealing with uncertainties
Edited by Markku Lehtonen, Pierre-Benoît Joly and Luis Aparicio

High Speed Rail and Sustainability
Decision-Making and the Political Economy of Investment
Edited by Blas Luis Pérez Henríquez and Elizabeth Deakin

Rawls and the Environmental Crisis
Dominic Welburn

A Green History of the Welfare State
Tony Fitzpatrick

The Governance of Green Urban Spaces in the EU
Social innovation and civil society
Judith Schicklinski

NGO Discourses in the Debate on Genetically Modified Crops
Ksenia Gerasimova

Sustainability in the Gulf
Challenges and Opportunities
Edited by Elie Azar and Mohamed Abdelraouf

Environmental Human Rights
A Political Theory Perspective
Edited by Markku Oksanen, Ashley Dodsworth and Selina O'Doherty

African Philosophy and Environmental Conservation
Edited by Jonathan O. Chimakonam

African Philosophy and Environmental Conservation

Edited by Jonathan O. Chimakonam

Routledge
Taylor & Francis Group
LONDON AND NEW YORK

First published 2018 by Routledge

2 Park Square, Milton Park, Abingdon, Oxfordshire OX14 4RN
52 Vanderbilt Avenue, New York, NY 10017

Routledge is an imprint of the Taylor & Francis Group, an informa business

First issued in paperback 2019

British Library Cataloguing-in-Publication Data
A catalogue record for this book is available from the British Library

Library of Congress Cataloging-in-Publication Data
A catalog record for this book has been requested

ISBN: 978-1-138-29713-5 (hbk)
ISBN: 978-0-367-26306-5 (pbk)

Typeset in Times New Roman
by Apex CoVantage, LLC

To all my African Philosophy students, undergraduate and postgraduate at the University of Calabar (2014–2017), for the intense engagements we had!

Contents

Figures

Preface

The idea to put together a volume on African environmental ethics first occurred to me in 2013. A certain species of ant (I have no idea what they are called) had invaded my front yard. A friend advised me to repel them with a certain chemical compound. I actually bought the chemical but changed my mind. Why kill these ants that mean no harm to me? Why decimate a whole ant colony that desires survival like us humans? I thought maybe that throwing the deadly chemical on the ants would be like a heartless dictator throwing a nuclear bomb on the whole of humanity. Obviously lacking the moral courage, I left them to live in my front yard. I wanted them off my yard but I did not want to kill them.

It must have been several weeks later when I woke up to see that the ants had killed a two-metre-long venomous snake. I was surprised because one rarely sees snakes in Calabar. It was a hot March weather and I had gone to sleep the previous night with the windows wide open. Look, I said; these ants have saved me from this snake. For the way they had devoured the snake, I nicknamed them tiger ants. But my tiger ants had not killed the aging snake because they had no conscience; they did it for thousands of them to survive. If they had language, the expression probably might not have been 'killed the snake' but something close to 'captured food.' I spared the lives of the ants; they saved mine. The old snake that would have died sooner anyway did not die for nothing. It had saved thousands of starving ants. Immediately, I was struck by a wonderful idea. It is all about 'relationship' or 'communion'!

Our forefathers here in Africa knew and taught the power and the value of relationship. Everything depends on something else. Our world is one big web characterised by relationship of mutual interconnection, interdependence and interrelation. This is the African way. It is not about which entity possessed reason or sentience or life; it is about the value different entities in nature contribute to render nature stable. On the basis of the above experience, I concluded therefore that African philosophy had a big task of showing humanity at large the path of honouring their relationships with other things in nature. Thus I was inspired to do a book of this nature.

This collection on African Philosophy and Environmental Conservation is the fruit of that little but intellectually rewarding encounter with the tiger ants. Philosophy, generally, compels us to honour our duties to humanity. What would be the value and relevance of African philosophy specifically if it could not address the challenges of the African place? I found to my chagrin that some big problems of our age, such as climate change, global warming, environmental degradation, women's issues, etc., were not receiving serious attention from African philosophers. What then is the point of the Great Debate and the struggle to affirm the existence of African philosophy, if the whole effort could not lead to a truly African paradigm of knowledge? The value of African philosophy in our time will not be measured by how many scientific papers are dedicated to its affirmation; it will rather be measured by the number of problems bedevilling Africans which it is able to confront. It is worrisome that, since the Great Debate, much of what African philosophy has done is either commentary on the ideas of Western philosophy or confrontation with the same. Very few works have been produced in African philosophy that ultimately engage with, and attempt to solve, problems facing the African peoples. This collection identifies the environmental challenge in its different shades and the contributors engage with the different manifestations of the problem.

In philosophy generally, African philosophy has been treated as peripheral, and this has attracted much criticism in the recent times. However, one finds that, even within African philosophy, there are peripheries. The environmental problem is one of them. Part of the motivation for this project on the environment is to relocate the problem of the environment accordingly, from the periphery to the centre of discourse. If a severe environmental crisis were to break out in our world, Africa certainly would be one of the most affected. It falls on the African philosopher, therefore, as the anonymous African thinkers once said, to go after the black goat while there is still daylight.

With the exception of three chapters, 3, 5 and 15, which are reprinted here with the kind permission of their authors and publishers, the rest were presented at the University of Calabar International Colloquium on "Marginalisation in African Philosophy: Women and the Environment," which took place from September 21 to 23, 2016. This is the first of the two volumes collecting the papers of the colloquium. While this volume is on the environment, the second volume is dedicated to papers that focus more on the theme of women. These two volumes are part of the projects which various African philosophers are engaged in for the progress of the discipline. I have always thought that a philosophical work of this stature should ultimately aim at opening new vistas of thought. I, therefore, look forward to the opening of new vistas of thought on the question of environmental conservation from the discussions in this volume. I have tried really hard to minimise errors in this volume, and while the authors are responsible for the views expressed in their respective contributions, I must accept all those errors pertaining to typesetting and other omissions.

Jonathan O. Chimakonam
Ogoja, Nigeria
May, 2017

Contributors

Ada Agada, philosopher and writer, has just finished writing his doctoral thesis. He holds a Master of Arts and Ph.D. in Philosophy from the University of Nigeria, Nsukka. He has published several articles in national and international journals and contributed to chapters in highly praised books. The peak of his research endeavour is the completion and publication of his most important work, *Existence and Consolation: Reinventing Ontology, Gnosis and Values in African Philosophy*. This groundbreaking work in African thought won a 2015 Outstanding Academic Title Award from *Choice Magazine*, one of the most valued recognitions an academic work can enjoy in the year of its publication in the United States. He is a member of the Conversational School of Philosophy (CSP).

Alexander Obiefoka Enukora Animalu was born on the 28th of August 1938 at Oba in Idemili South LGA of Anambra State, Nigeria. He studied at the University of Ibadan, Nigeria for his BSc and at the University of Cambridge, United Kingdom from 1962 to 1965, where he obtained MA (Cantab.) and Ph.D. (Mathematics) (Ibadan) in Theoretical Solid State Physics. His mainline Ph.D. research work, subsequently published in *Philosophical Magazine* became a citation classic by 1983, having been cited more than 729 times between 1965 and 2001. In 2013, the article was listed by Thompson Reuters (at number 13) among the Top 100 cited *Philosophical Magazine* articles of all time (between 1900 and 2010).

Professor Animalu's research at MIT led to his authoring a book entitled "Intermediate Quantum Theory of Crystalline Solids" by Prentice-Hall in 1977. An Indian edition was published in 1978 and was translated into Russian by the Russian Academy of Science in 1981, and the second edition (2014) was printed by SNAAP Press Ltd Enugu. He returned to Nigeria to become a Professor of Theoretical Solid State Physics in the University of Nigeria, Nsukka, in 1976, rising to become Head of Department of Physics, UNN in 1981 and 1994 and Dean, Faculty of the Physical Sciences, UNN (1987/88).

Professor Animalu was the 1990 Ahiajoku lecturer, the highest Igbo academic forum. He was appointed Director of the National Mathematical Center, Abuja (April 1999–January 2001). He retired in June 2001 as Emeritus Professor of Physics to become the founding Chairman/CEO of the International

Center for Basic Research. For his contributions, Professor Animalu received the 2000 Nigerian National Order of Merit (NNOM) Award for Basic Science (Nigeria's highest award for academic achievement). He became the President of Nigeria Academy of Science (the Apex scientific organisation in Nigeria) (2001/02), and in 2001, he was appointed a member of the Honorary Presidential Advisory Council on Science and Technology in Nigeria.

Professor Animalu is author of numerous scientific articles and more than 30 books in both the sciences and the humanities, including biographies of great men.

Kevin G. Behrens is a Senior Lecturer at the Steve Biko Centre for Bioethics, School of Clinical Medicine, Faculty of Health Sciences, University of the Witwatersrand (Wits), Johannesburg. He holds a doctorate in Public Philosophy and Ethics from the University of Johannesburg and a Masters in Applied Ethics for Professionals from Wits. Kevin is the academic co-ordinator of the Ph.D. programme in Bioethics and Health Law, and the Unit Head for the unit "Foundations of Bioethics" in the MSc (Med) in Bioethics and Health Law programme. His research interests lie in the area of Applied Ethics, in particular in Bioethics and Environmental Ethics. A major emphasis in his work is on applying sub-Saharan African moral philosophical notions to ethical questions. He has published widely in international and national journals and is on the editorial board of the *South African Journal of Bioethics and Law*. He holds a rating as an "Established Researcher" (C1), from the National Research Foundation.

Jonathan O. Chimakonam, Ph.D, is a Senior Lecturer at the Department of Philosophy, University of Calabar and a Research Fellow at the Department of Philosophy, University of Pretoria. His teaching and research interests include Logic, African Philosophy, Philosophy of Maths, Bioethics, Philosophy of Mind and Environmental Ethics. His articles have appeared in journals such as *Ultimate Reality and Meaning, Confluence, South African Journal of Philosophy, Thought and Practice, African Studies Quarterly, Ufahamu, Online Journal of Health Ethics, Dialogue and Humanism, Journal of Global Ethics, Journal of Pan African Studies, Phronimon, Polylog, Filosofia Theoretica*, etc. He is the author and editor of several books, including *Atuolu Omalu: Some Unanswered Questions in Contemporary African Philosophy* (New York Lanham: University Press of America, 2015). He has written many book chapters. He is the founder and convener of the elite African professional body, The Conversational School of Philosophy (CSP) and the founding editor of *Filosofia Theoretica: Journal of African Philosophy, Culture and Religions*. He is the winner of a Jens Jacobsen Research Award for Outstanding Research in Philosophy by the International Society for Universal Dialogue, conferred at the Polish Academy of Sciences, University of Warsaw, Poland (July, 2016). He is African philosophy Area Editor in the Internet Encyclopaedia of Philosophy. He propounded the theories of Conversational Thinking and Ezumezu Logic. Chimakonam has given several international conference lectures.

Francis Diawuo holds a Bachelor of Arts Degree in Integrated Community Development from the University for Development Studies, Wa-Campus, Ghana. He is currently a final-year student in Master of Philosophy (MPhil) in Environment and Resource Management at the same university. His MPhil thesis is on Women's Participation in Agroforestry Practice. His current research interests cover African Traditional Belief Systems and Natural Resource Conservation, Water and Sanitation issues. He also writes short articles on climate change and environmental issues for blog posts.

Ademola Kazeem Fayemi, Ph.D., teaches Philosophy at the University of Lagos, Nigeria. He is an Erasmus Mundus Fellow in Bioethics. His areas of competence include African philosophy, Socio-political philosophy, Bioethics in Africa, and Research ethics. Fayemi's publications have appeared in internationally indexed journals. He has read papers at international conferences in Europe, North America and West Africa.

Abdul Karim Issifu holds a Bachelor of Arts degree in Integrated Community Development and a Master of Philosophy degree in Peace and Development Studies from the University for Development Studies-Ghana (Wa Campus) and the Institute for Development Studies, University of Cape Coast-Ghana respectively. His previous publications include content related to indigenous peace building, chieftaincy and ethnic conflicts, and he has published in the *Journal of Global Peace and Conflict*, *Peace Studies Journal*, and the *International Journal of Corporate Strategy and Social Responsibility*.

Bruce Janz is Professor in the Department of Philosophy at the University of Central Florida, graduate faculty in the Texts and Technology Ph.D. program, and co-director of the Center for Humanities and Digital Research. He has been at UCF since 2003, and was chair of the Philosophy department from 2008 to 2013. Previously, he was at Augustana University College (now the Augustana Faculty of University of Alberta), in Alberta, Canada. His Ph.D. is from the University of Waterloo in Canada. He has taught in Canada, the US, Kenya, and South Africa. He is the author of *Philosophy in an African Place*, co-author of *A Phenomenological Analysis of Awe and Wonder: Towards a Non-Reductive Cognitive Science*, and editor of *Hermeneutics Place and Space*, along with other articles and chapters. His areas of research include contemporary African philosophy, theories of place/space, urban studies, environmental thought, contemporary European philosophy, digital humanities; interdisciplinarity, cultural studies, and philosophy of mysticism.

Workineh Kelbessa is an Associate Professor of Philosophy at Addis Ababa University, Ethiopia. He obtained his BA degree in Philosophy from Addis Ababa University, Ethiopia, his Master's in Politics of Alternative Development Strategies from the Institute of Social Studies, the Netherlands, and his Ph.D. in Environmental Ethics from Cardiff University, UK. He is the author of *Traditional Oromo Attitudes towards the Environment: An Argument for Environmentally Sound Development* (2001), and *Indigenous and Modern*

Environmental Ethics: A Study of the Indigenous Oromo Environmental Ethic and Modern Issues of Environment and Development (2011). He has also published many articles in refereed journals and chapters in books in the fields of environmental ethics, African philosophy, globalisation, philosophy of sex, and indigenous knowledge. His research interests include environmental philosophy, African philosophy, development ethics, globalisation, philosophy of love and sex, climate ethics and indigenous knowledge. He is a former Fellow of the Alexander von Humboldt Foundation, and a member of the UNESCO's World Commission on the Ethics of Scientific Knowledge and Technology.

Ralph Okechukwu Madu Born September 28, 1955, Rev Fr. Ralph Madu did his primary education at St. George's school in Falomo, Lagos. His secondary education was in St. Peter Claver seminary, Okpala. He obtained a B. Phil (1979) and a BD (1983) from Urban University, Rome. He also obtained an HND in Journalism from ESUT, Enugu. He holds a Master's degree and Ph.D. in Philosophy from the Catholic University of Leuven, Belgium, and also a BA in linguistics and MA in Religious Education from the same institution. Fr. Ralph received a Master's degree in Education Supervision and Administration and PD from Fordham University, New York, in 1992. 1992–1997, he lectured and served as head of the Department of Philosophy in Bigard Memorial Seminary, Enugu. From 1997–2006, he lectured at Seat of Wisdom Seminary, Owerri. He was for a total of 11 years the editor of the *Leader Newspaper* and Managing Director of Assumpta Press, Owerri. From 2006–2012, he worked as National Director of Social Communications in the Catholic Secretariat of Nigeria. Since 2012, he has been the Secretary General, Catholic Secretariat of Nigeria. He has worked for 12 years with the Christian Association of Nigeria, CAN, and is a member of the Governing Council of the Veritas University of Nigeria, Abuja, and the Catholic Institute of West Africa, CIWA, Port Harcourt. He has been a Catholic priest for 33 years.

Fainos Mangena, Ph.D., is a Professor of Ethics and Applied Ethics at the University of Zimbabwe. He is currently the acting chairperson of the Department of Religious Studies, Classics and Philosophy at the University of Zimbabwe, as well as the Chairperson of the Research Ethics Committee in the Faculty of Arts. He is also a member of the University of Zimbabwe Ethics Committee. His areas of research interest include ethics and its relationship with the environment, education, culture, gender and politics. He is a winner of the prestigious African Humanities Programme Post-doctoral Fellowship funded by the Carnegie Corporation administered by the American Council of Learned Societies (ACLS), and is currently the interim President of the Philosophical Society of Zimbabwe (PSZ). His most recent publications include: "Can Africana Women truly embrace Ecological Feminism?" in *Filosofia Theoretica: A Journal of African Philosophy, Culture and Religions* and "Restorative Justice's Deep Roots in Africa" in *South African Journal of Philosophy*, 34 (1), pp. 1–12, "Philosophy in the House of Stone: A Critical Review" in *South African Journal of Philosophy*, 34 (2), pp. 1–14, *Philosophy in African*

Traditions and Cultures Washington, D.C: The Council for Research in Values and Philosophy, pp. 1–173 and *Sounds of Life: Music, Identity and Politics in Zimbabwe* published by Cambridge Scholars Publishing, pp. 1–294 (co-edited with Ezra Chitando and Itai Muwati).

Thaddeus Metz is Distinguished Professor at the University of Johannesburg, where he is affiliated with the Department of Philosophy. Author of about 200 books, articles and chapters, he is particularly known for having analytically articulated an African moral theory, applied it to a variety of ethical and political controversies, contrasted it to Western and East Asian moral perspectives, and defended it as preferable to them. Some recent publications include: "Replacing Development: An Afro-communal Approach to Global Justice," *Philosophical Papers* (2017); "Values in China as Compared to Africa," *Philosophy East and West* (2017); and "How to Ground Animal Rights on African Values," *Journal of Animal Ethics* (2017).

Victor C. A. Nweke is a Ph.D. student at the Institute for Advanced Studies in the Humanities, University of Duisburg-Essen, Essen, Germany. His major research interests include logic, African philosophy, and cross-cultural relations. He has authored some articles and presented papers in these areas in some academic fora within and outside of Nigeria. His works mainly border on the innovative development and application of philosophic principles and systems from the African Place to confront the challenge of cross-cultural misunderstanding, peace-building and sustainable development in Africa and across the globe. He is a foundational member of the Conversational School of Philosophy (CSP) and was a Research Assistant to the Convenor of CSP, Jonathan Chimakonam Okeke, Ph.D. Victor sees himself as a "Conversationalist"; he therefore approaches, interrogates and converses with works and authors in other Philosophical Traditions from the ambience of African philosophy. He is also a member of academic bodies such as the Nigerian Philosophical Association (NPA), International Society for African Philosophy and Studies (ISAPS), Canadian Philosophical Association (CPA), Association for Informal Logic and Critical Thinking (AILACT), etc. Victor is currently working on a manuscript that projects African philosophy as a Conversational Tradition of Philosophy without racial segregation.

Ebunoluwa Olufemi Oduwole is the Deputy Vice-Chancellor, Olabisi Onabanjo University, Ago-Iwoye, Ogun State, Nigeria. She is a professor of African philosophy. She is a distinguished Fellow, Erasmus Mundus Bioethics Programme. Her research interests are mainly in African philosophy, Bioethics and Women studies, where she has published widely in both local and international outlets. Professor Oduwole has attended scholarly conferences in different parts of the world, such as Rio de Janerio in Brazil (2005) Windsor in Canada (2007), Leicester in the United Kingdom (2006), Leipzig in Germany (2009), Texas (2010), Johannesburg (2013), Austin (2016), among others.

Olatunji Alabi Oyeshile, Ph.D., is a professor of African Philosophy, Metaphysics and Existentialism in the Department of Philosophy, University of Ibadan. He has published extensively in books and journals of international repute, some of his publications include: *Reconciling the Self with the Other: An Existentialist Perspective on the Management of Ethnic Conflicts in Africa*. Ibadan: Hope Publications, 2005; Co-edited with J. Kenny, *The Idea of a Nigerian University: A Revisit*, Washington D.C: The Council for Research in Values and Philosophy, 2013; "Towards an African Concept of a Person: Person in Yoruba, Akan and Igbo thoughts," *Orita*, XXXIV/1–2; 2002; "Communal Values, Cultural Identity and the Challenge of Development in Contemporary Africa," *The Journal of Social, Political and Economic Studies*, Vol. 29, No. 3. Fall, 2004; "Corporate Existence and Individual's Freedom in African Communal Society: The Yoruba Example," *Ultimate Reality and Meaning Journal (URAM)*, Vol. 30, No. 4, 2007 and "Sen's Realization – Focused Notion of Justice and the Burden of Democratic Governance in African Societies," *Indian Journal of Human Development*, Vol. 5, No. 1, January, 2011. His research focus in recent times has been in the area of African Socio-Political Philosophy, where he has paid attention to the place of the individual within the community and also on how to resolve some perennial problems in the area of religion, politics and nationhood. He is a member of the Nigerian Philosophical Association (NPA), Nigerian Academy of Letters (NAL), International Society for Universal Dialogue (ISUD) And the Conversational School of Philosophy (CSP).

Angela Roothaan is a philosopher from the Netherlands. She earned her Ph.D. in philosophy in 1996, with a study on Spinoza's Theological-Political Treatise (published in Dutch). After holding several temporary jobs as a teacher and researcher, she gained tenure at the Free University Amsterdam, where she currently works. She consequently published books (also in Dutch) on Nature, in Ethics, on Truth, on Spirituality and on Ghosts/Spirits in Modernity, as well as numerous articles (in several languages) on the interconnected fields of anthropology, ethics, spirituality and political theory. Her most recent work focusses on questions concerning human and non-human animal relations, as well as relations between humans and plants and other natural phenomena. To investigate these questions, she mainly uses postcolonial as well as hermeneutic and deconstructive strategies of analysis. From 2013 on she regularly writes a blog, with the aim of getting philosophy out of the ivory tower: http://angela roothaan.wordpress.com

Jeff Unaegbu is a Principal Cinematographer in the Institute of African Studies, University of Nigeria, Nsukka, with research interests in social change, Paremiology, Paremiography and cinematography. He is the founder of Fellows of Inspiration, Research and Memory (FIRM, Nigeria). A prolific writer, seasoned editor and trained Mass Communicator, Unaegbu has authored four books, co-authored six books, co-edited four books, published numerous articles and produced 20 video documentaries.

Oladele Abiodun Balogun received his Ph. D in Philosophy from the University of Ibadan. He is a Professor of Philosophy at Olabisi Onabanjo University, Ogun State, Nigeria where he currently serves as the Dean of Faculty of Arts. His papers have appeared in international journals such as: African Identities 3 (2) (2005), Nordic Journal of African Studies 16(1) (2007), The Journal of Pan African Studies 2 (3) (2008), Philosophia: International Journal of Philosophy 38 (1) (2009), Thought and Practice: A Journal of the Philosophical Association of Kenya 1 (2) (2009) among other numerous scholarly journals. His research interests are in African Philosophy, Philosophy of Education, and Philosophy of Law. Balogun has attended international philosophical conferences in Ghana, Leicester (U.K.), South Africa, Korea, Singapore, Texas and Minnesota in America. When not engaged in deep thought and writing, he enjoys hanging out with his marvellous children, and lovely wife, Temitope. He is a member of the Conversational School of Philosophy (CSP).

Acknowledgements

This book has come a long way, yet not until we began to process it for publication did it occur to me how much of a debt of gratitude I owe different people. I thank the Vice Chancellor of University of Calabar, Professor Zana Akpagu, for the partial funding the University provided for the International Colloquium at which most of the papers in this volume were presented. I owe a basket of gratitude to Distinguished Professor Thaddeus Metz of the University of Johannesburg and Professor Louise d'Toit of Stellenbosch University, who assisted me greatly in organising the colloquium where most of the essays in this volume were presented. I thank Professor Metz specifically for providing financial support, which enabled some of the delegates to attend the colloquium and the University of Johannesburg, First Registrars, Professor Oladele Balogun, Professor Olatunji Oyeshile, Professor Eno Nta and Mr. Arinze U. Chizube of Airlotech Global Concept Ltd, for various levels of financial support during the colloquium. My special appreciation goes to members of the Conversational School of Philosophy – The Elite African Philosophy Circle, who worked tirelessly with me to organise and host the International Colloquium. Mention must be made of Dr. Mesembe Edet, Victor Nweke, Segun Samuel, Aribiah Attoe, Eloka Uduagwu, L. Uchenna Ogbonnaya, Umezurike Ezugwu, Amadioha Egwu, Charles Onyeanwuna and others. I thank also my undergraduate African philosophy students who lightened up the colloquium with their participation and made me really proud. I cannot forget the Head of Department, Professor K. A. Ojong and colleagues at the Department of Philosophy, as well as the Dean of Faculty of Arts, Professor Dorothy Oluwagbemi-Jacob, Professor Chris Ijiomah, Dr. Ephraim Ikegbu and others for their support during the colloquium. I thank in a special way the Director (Professor Chris Nwamuo) and Staff of Bassey Andah Institute for African and Asian Studies and the Director and Staff of the Directorate of Research and Quality Assurance, both of the University of Calabar, for providing us with a venue, support and facilities during the colloquium. Perhaps the best part of my appreciations should go to colleagues who honoured my invitation to come to Calabar from different countries and make their presentations, without which we would not have this volume on African Philosophy and Environmental Conservation. Finally, I thank some friends that assisted in one way or the other either during the colloquium or in the preparation of this volume, including Mr. Ayi Okon,

Mrs. Victoria Anna, Mr. Douglas Afu and a host of others. Finally, this volume contains three previously published articles and I heartily acknowledge the authors and the original publication venues of those essays:

Thaddeus Metz. "How to Ground Animal Rights on African Values: A Constructive Approach." This essay was first published in the *Journal of Animal Ethics* Vol. 7, No. 2 (2017), pp. 163–174 and is reprinted here with some minor changes and with permission from the University of Illinois Press.

Olatunji A. Oyeshile. "Transformation of Urban Space in South-West of Nigeria, 2011 to Present: Ethical Issues in Development and Aesthetics." This essay was first published in *Africa: Journal of Contemporary Issues*, Vol. 13, No. 13 (2015), PP. 35–50 and is reprinted here with minor adjustments and with the permission of the author and the publishers.

Francis Diawuo and Abdul Karim Issifu. "Exploring the African Traditional Belief Systems (Totems and Taboos) in Natural Resources Conservation and Management in Ghana." This essay was first published in the *Journal of Pan African Studies*, Vol. 8, No. 9 (2015), pp. 115–131 and is reprinted here with some minor changes and with the permission of the authors and the publishers.

In putting this volume together, the editor was supported and funded by Airlotech Global Concept Ltd., Nigeria, an environment-friendly company.

Introduction

Should African philosophers care about the sub-Saharan environment?

Why should African philosophers care about the African environment? Sub-Saharan Africa has one of the highest poverty levels in the world. Daily, millions of people depend, for example, on their forests and on the resources they can exploit from their environment for survival. Game hunting that requires bush burning, the need for firewood to cook their meals, wood logging, hide and skin, medications, etc., are some of the basic economic provisions people exploit from their environment, either for subsistence or for commercial purposes. When the government fails to guarantee their basic needs, people turn to their forests and the resources from their general environment for survival. It is clear that without these environmental resources, the poverty in these African communities will reach alarming levels. In the presence of these critical situations, why should an African philosopher care more for the environment? And on what moral grounds should he counsel the people against the exploitation of their environment?

It is on the basis of a dilemma such as the one above that we may begin to reflect on the relationship between African philosophy and environmental conservation. Of course, ignorance is always the main problem with regard to the abuse of the environment in some places, particularly in the Western world. Think of Donald Trump's mantra against the climate change campaign. The new US president and some others who think in this direction are convinced that the notion of climate change is a farce. He wants to drill the oil reserve in the arctic region, build heavy industries that threaten coral reefs and aquatic life and destroy forests and parks in order to build oil pipelines. In this and other similar regards, it is ignorance that is culpable. But think of sub-Saharan Africa, where some of the major threats to the environment are not the heavy industries and chemical plants but the activities of individual citizens living in various communities. The only way for these poverty-threatened people to survive is by direct massive exploitation of their environment. In this context, therefore, poverty rather than ignorance is the main danger. Everything changes when the paradigm shifts from ignorance to poverty. Poverty, it should be observed, even metaphorically, devours, and that is what is happening to the environment in sub-Saharan Africa.

So, yes! African philosophy should care about the African environment; and the African philosopher has a role to play in protecting the African environment that is under serious threat. Take, for example, the endangered Cross River forests in Nigeria. The Cross River State is in danger of losing their forest, which is home to an impressive portion of world's biodiversity. The state has an area of 20,156 km^2 (7,782 sq mi) and a population of 3,337,517. It is home to the Cross River National Park, a world Biodiversity hotspot covering a total area of about 4,000 km^2. This park has two divisions, namely, Oban Hills and Okwangwo. In the North and Central parts, there are mainly tropical rainforests, while in the coastal zones there are mangrove swamps. Some portions of the park are categorised under the Guinea-Congolian region with a closed canopy and trees that reach up to 50 meters in height. There are vast numbers of identified animal species in the park, even though a comprehensive study is yet to be carried out to identify all the species in the park. However, these vast forests have come under threat in recent years through illegal logging, poaching, burning, felling trees for firewood, etc. These are the activities of ordinary Cross Riverians who struggle to survive under an intense weight of poverty and lack of basic social goods. Thus, the big question: why should we protect the environment? Are we, the present-day Cross Riverians, obligated to future generations in the use of our forests? If, as some argue, we must protect the environment because we owe a moral duty to posterity, the question that looms is: what has posterity done for me/us? Moral dilemmas like this resonate when the matter is about poverty and survival rather than mere ignorance.

It is, therefore, the burden of the African philosopher to convince Cross Riverians, for example, that they owe moral obligations to the future generations with regard to the use of their environment. This may seem trivial and, indeed, of little importance to the men and women whose daily means of subsistence appear to depend on these forests, one way or the other. But the bitter truth is: yes, we owe posterity a mountain of obligations with regard to how we use our natural environment and our forests specifically.

In the unlikelihood of a global plague that will wipe out the human population, we can agree that future Cross Riverians will come and will continue to come for many more years, if not centuries. Before the present generation came, many generations have lived and walked on this land. They did not plunder and obliterate the forests; hence, we find the forests useful in our time. To answer the question whether we owe any moral obligations to future generations in our use of these forests, let us imagine that the topography of Cross River State today is like the Sahara Desert. Surely, we can imagine the heat and the thirst and the famine. Then someone comes along to show us pictures of how Cross River state used to be like; he shows us the lush green vegetation from Odukpani to Ogoja, the rings of fresh water rivers from Ikom to Calabar, the basket of freshwater fish along Lagos street and the mountain of harvests from various hinterlands. Would we not feel betrayed by our ancestors (the previous generations) who destroyed the forests and desiccated the land? Indeed, how disgusting their memory must seem to us! And, indeed, how disgusting our memory will be to future generations should we destroy the Cross River forests.

It is so easy for some of us, out of sickening selfishness and embarrassing short-sightedness, to argue that all that matters is today and today alone. They say that the future generations do not matter and should not matter to us because they do not exist; how can we be morally obligated to people who do not exist? They say that the survival of the present generation is paramount; that if we starve and deny ourselves the luxury of our environment just to bequeath it to the future generations, how can we tell that the so-called future generation will be like us and have the same needs and values as we presently do?[1] And then, from a selfish standpoint, they wonder why we should be obligated to care about the future people who we do not even know and will never know. They ask, where is the sense in denying ourselves the luxury of our environment for a future generation that does not and will never care about us? Indeed, they ask quite shamelessly, what has the future generation ever done for me?[2]

But one may ask these types of people; what have you ever done for the previous generation? What did you do for them to deserve the inheritance of these forests? I am not feigning ignorance of the importance of firewood to thousands of homes in Cross River State and other places in the sub-Sahara; neither am I ignorant of many family bread-winners whose income depends on logging, but the fact remains that there are viable alternatives. If we allow these people to plunder these forests, in less than five years, the Cross River forests will be gone. The question is; will those families not continue to cook their meals? And will those who make income from logging not continue to support their families? So, it is really not a question of 'the only option' that makes people plunder the environments in the sub-Sahara like the Cross River forests. Shamefully, it may be a question of the 'easy option.' In issues of the natural environment, if the options are such that the easy one will serve our luxury but will lead us to disaster, do we not have a moral obligation to choose a slightly difficult one that offers stability and survival of our species? I certainly think we do!

John Trumbull, an American poet, renowned in his time, asked some 200 years ago, why do we behave

> . . . as though there were a tie
> And Obligation to posterity.
> We get them, bear them, breed, and nurse:
> What has posterity done for us?

Well, it is important to tell those of us in the Cross River State of Nigeria and other places in the sub-Sahara who may think along this line that it is not so much a matter of what posterity has done for us that should determine our attitude toward the environment; it is rather what we can do for posterity. In matters concerning the protection of our common home – the earth – and the survival of our human species, the debt of moral obligation is one-directional; the previous generation owes us, we owe the future generation, and they owe the generation next and so on. If we have found the earth a good place, then we should conserve it to be good for posterity as well. And if we have found life a lovely gift, then we should do everything possible to pass it on and allow the future generation to receive and enjoy this gift of life. This entails moral obligation. It is the present generation that owes the future generation. It is in our power to end it all. We can decide to build and detonate a nuclear weapon that can sink the earth and obliterate life, but why should we make such a choice? Obviously, there is no good reason for that. But there are many good reasons why we should pass on the baton to the future generation, including the conservation of our common home and the protection of our species.[3]

It may therefore be asked, why does species protection matter?[4] Species protection is the concern of biodiversity. The Cross River forests, for example, are an important biodiversity hotspot in the world. As the world-renowned leader in biodiversity studies Edward Wilson stated in *The Diversity of Life*, "biodiversity is our most valuable but least appreciated resource."[5] The earth is able to support all forms of life by the mechanism of biodiversity. Different species serve as food or life saving medicines or provide conditions necessary for the survival of other species. Without biodiversity, even the human species will not survive. It, therefore, becomes a moral duty for us to protect the African forests which are home to many species of living things.

As African philosophers, it is our duty to pass this message to the people who may not be so privileged as to be in possession of this knowledge. It is a battle against ignorance and poverty, a battle to be won with the power of knowledge rather than with the force of arms. The earth is our common home. The future generations are entitled to it as much as we are entitled to it. The earth without its forests will be like the planet Mars, which scientists tell us cannot support life. The battle to protect and conserve the sub-Saharan environment has to be led by African philosophers who have the tools to fight both ignorance and poverty and diffuse the dilemma of survival and conservation.

In this collection, African philosophers from different parts of the world theorise and offer insightful opinions on how to conserve the African environment. The chapters in this book are papers presented at an International Colloquium held at the University of Calabar, Cross River State, Nigeria in September 21–23, 2016. The collection is rich and unique because it is, to my knowledge, the first volume to address African philosophy's role in the conservation of the sub-Saharan environment.

In chapter one, Bruce Janz employs two concepts: peripherality and non-philosophy to investigate the themes of womanist philosophy and environmental philosophy, respectively, in the field of African philosophy. He argues quite interestingly that these two themes, regrettably, have been treated as peripheral and of little interest in African philosophy. If African philosophy were to add another meaningful layer to its discourse, Janz insists that questions about women and the environment should be given significant attention by its practitioners. This is an eye-opening piece that draws attention to how African philosophy, which has often served as peripheral and as non-philosophy to Western thought, harbours its own periphery and non-philosophy.

Alexander Animalu and Jeff Unaegbu, in chapter two, dig into the Igbo worldview to unearth the semblance of the well-known Greek Gaia and make a case against the marginalisation of women and the environment in African philosophy.

This is a must-read for those interested in the ideas inherent in African traditional thought.

Thaddeus Metz in chapter three expounds his theory of 'African Modal Relationalism,' which is a theory of animal rights with a sub-Saharan pedigree. He defends it from several criticisms raised by Kai Horsthemke in his book *Animals and African Ethics*. This is an interesting piece that shows not only how animal rights might be defended in African philosophy but the global viability of this approach in environmental ethics.

In chapter four, Kevin Behrens argues that there are notions in African thought that can enrich our discourse about obligations to preserve biodiversity. He shows that a sub-Saharan conception of moral considerability, grounded in the notion of an interconnected "web of life," is able to accord moral standing to species *qua* species. Additionally, he gives an account of an African conceptualisation of our moral duties to future generations that entails a strong obligation to preserve biodiversity for posterity. The work is an exciting piece that counters the idea that the African worldview is notoriously anthropocentric.

Olatunji Oyeshile in chapter five writes about the human cost which environmental renewal might have in developing nations like Nigeria. He argues that development is meaningful only when it is not only about aesthetics but also about people who are the ultimate beneficiary. He recommends that states in Nigeria should balance urban renewal with the survival of the people. He posits that a gradualist approach that provides a viable alternative source of livelihood is preferred to the radical transformation of urban space, given the poverty level of the people, if sustainable development is to be realised. This is a curious anthropocentric piece.

In chapter six, Ebunoluwa Oduwole and Ademola Kazeem Fayemi expose and interrogate the relationship between humans and non-human animals in traditional Yorùbá African culture in order to cogently frame the theoretical leaning of the Yorùbá peoples' worldview in the ongoing global discourse on the ethics of animal relations through hermeneutic exploration of the Yorùbá oral tradition. They conclude that the Yorùbá are advocates of ontological animal care ethics; in this rational framing, religious beliefs, taboos, folktales, proverbs and moonlight poetry are essentially instrumental in serving as restraints to the mistreatment of animals. This is a rich philosophical exploration of animal care ethics in Yoruba culture.

Workineh Kelbessa in chapter seven explores the contribution of African women to environmental protection by taking some examples from different countries. He argues that, although women held positions as chiefs, had their own organisations and participated in male councils in some cultural groups in pre-colonial Africa, women in other cultural groups were subject to different aspects of subordination because of customary laws. Colonisation for him consolidated the master mentality that denigrated women by altering the customary gender division of labour. He suggests that women should empower themselves and challenge patriarchal values that perpetuate their exploitation. African governments, he argues, should also pay greater attention to women's roles, knowledge, perspectives, and needs in order to solve environmental problems. This is an exciting piece on the importance of women's roles in environmental protection in Africa.

In chapter eight, Ademola Kazeem Fayemi and Oladele Abiodun Balogun employ the tool of conversational thinking to interrogate the perspectives of three female Nigerian philosophers on the idea of African environmental ethics. The philosophers are Sophie Oluwole, Philomena Ojomo and Pat Agboro. They discover that the common identities that straddle across the writings of these three philosophers on the topic are female authorship and the geographical criterion of Africanness of a thought. They found that neither of the positions of the three selected scholars shares a feminist bias for environmental concerns, nor underscores an appreciation of the roles and constraints of women in sustainable environmental management, be it in Africa or elsewhere. Using the African eco-activist and eco-feminist bias of Wangari Maathai as motivation, the authors then call for a womanist bias in African environmental ethical discourses. They argue that there is an important nexus between women's empowerment and a good environment, especially in the African context. For a holistic orientation in African environmental ethics, which will serve as pivot for effective and sustainable management of the environment in 21st-century Africa, they conclude that the metaphysical must reinforce not only the ethical, womanist dimension, but also non-philosophical variables such as law, technology, education and eco-activism. Again, this is an exciting read for all.

Jonathan O. Chimakonam in chapter nine offers an articulation of a new account of the ecosystem called ohanife using the underexplored African notion of 'relationship' found in many places in sub-Saharan regional cultures and embedded in the Igbo notion of ezi n'ulo, which is emblematic of a relationship of interdependence and complementarity. He argues that the ecosystem can be conceived as a community and a common home to all there is. In it, he argues, every entity, both human and non-human, matters. Everything is connected to this community, called ecosystem, and through it; everything is connected to everything else. He develops and deploys the ideas of complementary and conversational curves to demonstrate that in the ecosystem, everything, whether human or non-human, stands equal to others and represents a necessary link in the network of existents and that the rupture of this link might have unknown consequences. This is an original African contribution to the discourse on ecosystem.

In chapter ten, Angela Roothaan looks into the case study of the relations of human communities to trees, especially in West-African traditions, and asks how these contribute to a richer understanding of the environment. She argues that what is of special interest in African environmental issues is that the meaning of trees is contested in clashes between modernisation movements, religious conversion movements, secular conservationalist, and traditionalist movements. She argues further that in investigating the content of the contested – a rich hermeneutics of trees – can contribute to an understanding of the environment that could create more sustainable relations of humans to their environment. This is a rich piece in African environmental ethics with specific focus on trees.

Victor Nweke in chapter eleven addresses the challenge of global warming from the perspective of some theories in African philosophy as an academic discipline. He employs relevant principles and theories articulated by Innocent Asouzu, Jonathan Chimakonam and Thaddeus Metz to explain the basic cause(s) and possible ways to tame the tide of global warming. He shows the relationship between the theories examined and their prospects in relation to understanding and addressing the challenge of global warming. He argues that each of the theories examined contains plausible ideas that can be creatively harnessed to provide a philosophical framework for understanding and reversing the tide of global warming today. This is an interesting piece that is able to use the theoretical resources in African philosophy to make a case for environmental conservation.

In chapter twelve, Ralph Okechukwu Madu discusses ecological crisis from both the religious and the African perspectives. He argues that today, ecological crisis is proving to be one of the most daunting challenges, and perhaps the most arduous one that humans have faced since the beginning of life. He employs Pope Francis's recently published *Laudato Si* to re-invigorate discussion on the importance of addressing the ecological crisis. The crisis, he argues, is about our planet earth, where the survival of the biotic community, including human life, is increasingly under threat, caused by a weak appreciation of the interrelatedness of all existents as well as their interdependence. This is a fine religious, yet pragmatic, perspective on African environmental ethics.

Ada Agada in chapter thirteen presents a variant of the anthropocentric model of the environment, dubbed moderate anthropocentrism, as a viable environmental model for Nigeria and Africa. With references to empirical data, he establishes the negative reality of climate change in Nigeria. He further shows that it is established that traditional Nigerian (African) conservative practices are rooted in the African holistic view of the universe. He argues that both anthropocentric and non-anthropocentric models are human epistemological windows on the world, which validate the human-centredness of the anthropocentric model. He concludes that any attempt to exhibit the intrinsic value of non-human living organisms and non-sentient things must appeal to panpsychism for the conditions of its possibility if we are to go anywhere near the notion of 'fundamental rights' of the environment. This is an interesting contribution to African environmental ethics.

In chapter fourteen, Fainos Mangena argues that the environmental crisis which Zimbabwe faces is a result of the failure by the government of Zimbabwe and its people to harness hunhu/ubuntu ethics to live in harmony with nature. He claims that the hunhu/ubuntu ethic remains relevant to the discourse on environmental philosophy in Africa, despite some criticisms. This is a fine piece on environmental ethics from the perspective of ubuntu thought.

Concluding the volume, Francis Diawuo and Abdul Karim Issifu in chapter fifteen attempt to discover how taboos and totems as belief systems have been used successfully to promote natural resources conservation and management in the Sankana and Tongo-Tengzuk communities of Ghana. Using primary and secondary data sources, they observed that the communities have served as classic examples of successful traditional natural resources conservation and management through totems and taboos. On this basis, the authors recommend a holistic approach involving the government through the Ghana Tourism Authority, civil society organisations, and religious leaders to encourage people to apply by-laws and other traditional belief mediums that link human existence to the natural environment in the conservation efforts in Ghana. This piece shows just how the traditional African belief systems played and can still play vital roles in environmental conservation efforts in Ghana and elsewhere in Africa today.

On the whole, the contributors to this volume who are specialists in the African philosophy tradition have exerted enormous intellectual energy to excavate, construct, reconstruct and propound ideas in African philosophy on how to conserve the environment and its resources in the African place and beyond. One common theme that runs across the chapters is the idea of 'relationship' not just among humans but between humans and non-humans in the ecosystem. It is, therefore, suggested that in the ecosystem, everything, whether human or non-human, stands equal to others and represents a necessary link in the network of existents. Arguably, the general idea is that the rupture of this link among all the entities in nature might have unknown consequences. I invite scholars in philosophy, environmental sciences, cultural studies and the curious minds out there to this feast of African philosophy.

Notes

1 See Golding, M. (1972). Obligations to Future Generations. Reprinted in Pojman, L. P. (1994). *Environmental Ethics: Readings in Theory and Application* (pp. 225–230). Boston, MA: Jones and Bartlett Publishers. Golding argues that we do not owe any moral obligations to future generations because they do not meet his criteria (i.e., being recognised as part of our moral community and having the same values system as we do) for attracting moral duties from us.
2 See Pojman, L. P. (1994). *Environmental Ethics: Readings in Theory and Application* (p. 216). Boston, MA: Jones and Bartlett Publishers.
3 See Kavka, G. (1978). The Futurity Problem. In R. I. Sikora and B. Barry (Eds.), *Obligations to Future Generations*. Philadelphia, PA: Temple University Press. Kavka argues that because future people will have needs similar to ours as occupants of the earth planet and will be morally equal to us as members of the same species, we have a duty to them.
4 See Russow, L-M. (1981). Why Do Species Matter? Reprinted in Pojman, L. P. (1994). *Environmental Ethics: Readings in Theory and Application* (pp. 158–164). Boston, MA: Jones and Bartlett Publishers. Lilly-Marlene was one of those to critically address this question.
5 See Wilson, E. O. (1999). *The Diversity of Life*. New York: W. W. Norton and Company.

1 Peripherality and non-philosophy in African philosophy

Womanist philosophy, environmental philosophy and other provocations

Bruce B. Janz

Introduction

In this chapter, I want to think about two concepts: peripherality and non-philosophy. The first of these figured in my book, *Philosophy in an African Place* (Janz, 2009a), and designated the concepts and practices that stood in our peripheral rational vision. The specific context there was to look at the nature of tradition, and to provide an account of how we engage a concept, one which allowed us to think about the moments of tradition which necessarily exist in all rational action at a societal level. These concepts "in the periphery" scaffolded our intentional consciousness, that is, enabled our rational gaze to be rational by being related to but erased from our attention. Just as our sensory vision does not simply focus on a single thing, but focuses on things in the context of a set of other sensory experiences, both visual and otherwise, and as an experience of an embodied subject in a world, moving through that world and relying on it for the coherence of our object of attention, so too do our philosophical concepts reside in a larger structure of inclusion and exclusion.

The second concept is one which has received only a very specific and limited kind of attention in African philosophy. African philosophy itself has often found itself in the position of being the non-philosophy of others. Historically, that is certainly true, and for many philosophers the operative question still remains "Is there an African philosophy?" which amounts to assuming that it does not exist and the onus to prove it is on the one claiming that there is. There are many ways of calling African philosophy a non-philosophy without actually using those words. It might be a knowledge system, or a set of beliefs, or an ethnophilosophy, or a tradition, or a body of wisdom. It might be venerated as timeless or profound or perennial, but none of these are the same as a philosophy.

But this is not the non-philosophy I had in mind. I am interested in African philosophy's own non-philosophy. Few have addressed this. Theophilus Okere is one who has. In *African Philosophy: A Historico-Hermeneutical Investigation of the Condition of its Possibility* (1983), he has a chapter called "Philosophy and Non-Philosophy: Lessons from the History of Philosophy." His focus in that chapter, though, is almost exclusively on philosophy outside of Africa. The purpose of the chapter is to test his theory about how to philosophise, which is, as he puts it, "an interpretive commentary on reality against the guiding, determining, and suggestive background of a culture." Apart from Okere and those who have commented on his work (and a brief remark in passing by Mogobe Ramose), it is difficult to find anyone who has explicitly addressed African philosophy's own non-philosophy. The plan here, then, will be to think about that. We will unpack Okere's notion of non-philosophy and suggest an alternative understanding, based on something other than the implicit impulse when trying to include previously ignored areas of philosophy, which is the idea of the expanding circle. Including womanist and environmental philosophy is not analogous to Okere's approach to non-philosophy, which assumes that we need to find the initial content of philosophy in areas outside of the philosophical tradition (mainly religion and culture). We will see instead that thinking of non-philosophy in another manner than Okere does opens up the possibility of thinking womanist and environmental thought as more than just a lack in current philosophy that needs to be filled.

Peripherality and non-philosophy are necessary to think about womanist philosophy and environmental philosophy in Africa. The tendency is to recognise the reality that neither area has had very much attention, and then ask how we might expand the range of African philosophy to include these. This move of expansion is the one I want to question, and I would like to offer something else instead, that I think will more credibly enable African philosophy to engage with new areas of questioning. In other words, I am interested in thinking about the ways that African philosophy might and should encounter new areas, as well as ignored and avoided areas of inquiry. I want to resist supporting an implicit hierarchy, in which what is done now in any philosophical tradition is "real" philosophy, and what has not been addressed is peripheral philosophy. Both peripherality and non-philosophy should lead us to reconsider those received hierarchies.

The expanding circle

I use the image of the expanding circle in part because it has been used outside of African philosophy to think about how to have an ethical theory that can include the environment. Peter Singer is the best-known proponent of this. While I am not aware of the expanding circle image being explicitly used in African philosophy, I would argue that the move is used implicitly and is a familiar one. If we have an existing body of research in this area, and if that work has theorised things like ontology and subjectivity in a manner that is presented as genderless but which in fact has overwhelmingly been theorised by males, and which furthermore has focused on human relations, existence, and community, then the question of what to do about these blind spots seems to naturally raise the possibility of enlarging the circle.

The mode of questioning may not even be about gender or some other example of a marginalised area. So, for example, Odera Oruka's method of questioning the sages, at least early on, mostly consisted of asking them about concepts common in the Western philosophical canon, such as; What is truth? What is the good? To their credit, some of the sages did more than just try to answer those questions.

They answered different questions, ones which did not assume a set of categories that did not easily fit their experience. Oruka's interviews turn out to have philosophical merit, but sometimes in spite of his questions rather than because of them.

The point is that we use this model regularly, both within African philosophy and outside of it, whether we use the metaphor or not. At best, it is intended as a move of inclusion, indeed, generosity, and at worst it is patronising, as if those who have historically claimed the mantle of philosophy are secure in their place, and because of that security can afford to be generous to those who want to broaden the scope of philosophy. This is the model that I want to question, and to question it means asking about the border between philosophy and non-philosophy.

In a sense, I want to take a step behind the question of the place of women or questions about the environment in African philosophy, to think about how these can be encountered at all in a manner other than this model of expanded inclusion. It is a model which, in Europe at least, grows out of the questions of its place and time, out of the ideal of universality and universalism, out of a series of threats represented by very well-entrenched versions of non-philosophy in the Europe of the 18th century. Reason drew a bright line between itself and its others, and while the primary target of that action was the "superstition" that was seen to exist in religious, political, and cultural structures, the secondary targets and the collateral damage were everything that was not "Rational" as defined by those writing the treatises and encyclopaedias. Peter Park has done an excellent job of charting this history in his recent book *Africa, Asia, and the History of Philosophy: Racism in the Formation of the Philosophical Canon, 1780–1830* (2013). It is the activity of drawing this bright line that I want to question, and suggest that within an African context there are other options for imagining the relationship between philosophy and non-philosophy.

In a sense, I am speaking here in favour of the non-philosophical aspects of the places in which we find ourselves and which support and bear our thinking. To the extent that philosophy has the impulse to master its places, to subject them to its rational gaze and render their subtleties intelligible, we end up losing place. Yes, place is always something that is engaged, and that includes rational engagement. But it is not just that. The expanding circle model of philosophical engagement is, I think, that attempt to master new places rather than to encounter them and risk being changed by them. It assumes that the tools and methods of philosophy are already developed and honed, and areas such as women's experience or the environment are simply new spaces in which to exercise these well-established tools and methods.

The result of rethinking how we deal with non-philosophy will be directly relevant to how we deal with the contributions of women and the nature of the environment in African philosophy. Instead of looking for increasingly abstract principles that would allow us to extend the circle to new areas, we instead look for the ways that reason, in its broadest sense, is played out beyond our immediate sphere of attention. I have argued for exactly this in a previous essay in regards to the environment (Janz, 2009b). Instead of starting with abstract theories of African personhood or African ethics or African metaphysics, which are presented without any gendered inflection but which nonetheless generalise African concepts and knowledge from the overwhelmingly male practitioners of African philosophy, and then just adjusting them slightly to be "inclusive" or "sensitive" to gendered experience, what would it look like for epistemologies of gender to be the starting point, within an African ecology of concepts? Instead of developing an ethics or epistemology first, and then "applying" it or making it "practical" in the specific areas of the environment or gender, what would it mean to question the very foundations of African thought?

And are the environment and gender the only two challenges to the existing structures of African thought, or are there others? Are there challenges that are uniquely African? What would it mean, for instance, to re-theorise ancestors, not as just the basis for a moral order and a means to critique modernity through traditional categories, but instead to think of ancestors as paths that have been walked before, paths that might be honoured by being walked again, but might also be honoured by striking out anew in the same spirit of creativity as they showed? What would it mean to rethink the collective nature of African society, sometimes understood as Ubuntu? That too might be seen as non-philosophy, at least in the sense that it is rooted in cultural experience and praxis before it is theorised. So, what might it look like to re-theorise?

The point is that questions of gender and the environment are but two instances of the creative potential of non-philosophy in African philosophy. Simply thinking in terms of extending the circle cuts us off from seeing that creative potential. We need another way to proceed.

Peripherality

As a category or concept, it is tempting to think of the peripheral as the secondary, the less-important, the optional. To the extent that we think peripherality means something like that, it would follow that womanist philosophy and environmental philosophy seem peripheral to African philosophy at a descriptive level. In other words, the bulk of the work in the field has not directly taken up concerns that might arise for these two. It would further follow that this is a situation that ought to be corrected. We ought to be paying attention to these areas.

At one level, this line of reasoning is fine, but it does raise some nagging questions. Why is it that they have been avoided or ignored to this point? Is it that there has been little or no voice from women in African philosophy, and no one to represent the environment? That would certainly be true, but that does not entirely answer the question. It seems intuitively obvious that these areas ought to be part of African philosophy, but then again it evidently seemed intuitively obvious decades ago to our earlier colleagues that these questions were not all that significant. So the question of voice is surely part of this, as we ask the obvious question: "Significant to whom?" And yet, if we simply proceed in that manner, we treat philosophy as an enterprise in which each identity must be represented. Indeed it must – but what of the identities we have not yet anticipated, which are not yet at the table, or which cannot speak for themselves? The problem, in other words, is not whether we ought to have included womanist and environmental thought from the beginning, because we should have. The question is, how do we determine what ought to be included when there is no one to speak for it? And, further, how can one speak for something else (e.g., the environment) when we can have at best a passing familiarity with it, no matter how hard we study it, and it is at any rate a construct, a way of speaking about a host of different entities, processes, and histories under one term? And, finally, how do we make possible a new line of flight, a new set of questions and room for thought in philosophy, that is not simply predicated on fixing past omissions but on finding new ways of spurring thought?

So, while we can acknowledge the past peripherality of these areas, in the sense just mentioned, my use here will be different than that. I am less interested in the idea of what has been marginalised in the past and more interested in the fact that any creation of concepts will necessarily mean two things – first, we will have to pick out something from among a wide range of possible concepts, which we regard as adequate to the place in question and might possibly be activated by our cognitive scaffolding, and, second, we will have to be aware of the ways that concepts continually depend on that non-intentional frame for their vitality, and they are constantly renegotiated as that frame shifts for a variety of reasons. In other words, the peripherality here is that of the movement of our concepts into the position of prominence, and the continued maintenance of that prominence. To see how this works, we can think about our other concept – non-philosophy.

On non-philosophy

Being the non-philosophy for other traditions, as African philosophy has been, gives a unique perspective when thinking about the edges of one's own tradition. The mistakes of Western thought are generally obvious to those outside of it. There is an air of smugness, an Enlightenment sense of certainty and entitlement which looks down on any tradition outside of its own. There is a kind of erasure of the source of ideas – once they enter into the Great Conversation, it no longer matters whether the ideas originated in Africa or somewhere else. They are now "universal" ideas, and everyone henceforth assumes that their provenance came through the West. There is historically little self-critical attitude about Western philosophy's own categories and methods when coming into contact with its own non-philosophy. There is, at best, an attitude of expanding the circle, in the sense of finding ways to extend existing Western patterns of thought to new areas.

These features of Western philosophy have meant that even when there are real contributions from outside of the centres of philosophical power and influence, they have not fundamentally questioned the practices of philosophy. That does not mean that there have been no questions – one might see the division between analytic and continental philosophy as being about each side's questions about the other side's categories and methods. The result, for many, was a chasm, rather than the ability to be self-critical. This can also be seen in areas such as feminist philosophy and environmental philosophy in the West. For each (and we might give other examples here as well), there was a sense that they were tangential to the primary mission. Words such as "applied" or "interdisciplinary" were used, and, for many, were signals that those who engaged in these things were not really doing hard-core philosophy, but something less. The potential for new questioning was neutralised, and the opportunity for real self-reflection was lost, or at least deferred for years.

As African philosophy considers the areas it has overlooked, it can learn both from Western philosophy's intransigence in the face of new currents of thought, and also its own status as non-philosophy in the opinion of the vast majority of Western philosophers. Merely expanding the circle, a "me-too" approach, leaves intact the existing practices of philosophy. It is an understandable first step. It is what we saw with Mary Wollstonecraft's response to Jean-Jacques Rousseau's *Emile* – a plea for women to also be included in an Enlightenment plan of education. This was necessary, but it was also necessary to go beyond this to change, or at least offer alternatives to, the way we think about philosophy.

Non-philosophy, I will argue, is both the limit condition of philosophy as well as its creative spark. What is not subject to rational inquiry nevertheless exists, at least as a barrier to thought but more positively as a question to thought, as a mystery to be solved. As I have argued before, philosophy is ideally not grounded on sets of propositions to be defended but on questions that open up new areas of inquiry, new problematics.

There are various versions of non-philosophy we might imagine. Mogobe Ramose, following Enrique Dussel, sees philosophy as being about the non-philosophical, in the sense of a reflection on reality, specifically its own reality (Ramose, 2015, pp. 551–552). Theophilus Okere sees culture or religion as the proper non-philosophy in the sense of the "raw" source of material for philosophical reflection, but we could easily imagine other possible sources of such raw material. Sensation, for instance, might be thought as a non-philosophy in the sense that it is a potentiality waiting to be theorised. Likewise, common sense might be seen as wisdom which has not gone through the formal structures of philosophy. Someone might want to see nature as a non-philosophy, something pre-existing humans and waiting to be thought (or perhaps resisting thought).

Or we might locate non-philosophy elsewhere than pre-philosophy. If philosophy is located in its textual traditions, for instance, we can also recognise that there are clearly many kinds of non-philosophical texts. If it is located in its questions, there are many non-philosophical questions. There are, in other words, other disciplines which, despite the early modern Western tendency to classify all disciplined knowledge as either "natural philosophy" and "moral philosophy," are not at this point reducible to philosophy. Another version of non-philosophy would be Francois Laruelle's, which is meant to counter the self-authorisation of philosophy as the only or the pre-eminent mode of thought (Laruelle, 1999, 2010; Maoilearca, 2015). And, finally, we could look to another kind of non-philosophy: wonder. Wonder is that which spurs philosophy, and as such must precede it. It is an ontological or cognitive condition which makes possible and motivates philosophy (for more, see Gallagher et al., 2015, pp. 115–129).

These versions of non-philosophy are clearly not all of the same sort. Some assume that non-philosophy is the material source of philosophy; some assume that it is the motivation, some of these are simply alternative forms of disciplinary knowledge, and some might be, for lack of a better term, the antidote to philosophy.

But our point of reference here will be Theophilus Okere's chapter on non-philosophy (1983, pp. 81–113). As already indicated, the chapter looks like it has little to do with African philosophy as such. He works mainly from Hegel and Heidegger, and finishes with some brief comments on the non-philosophy that might be found in Plato (via F. M. Cornford). And yet we can see his argument as one which very much addresses a central question of African philosophy, which is how one might go about doing philosophy in Africa. His meditation on Hegel and Heidegger is meant to give us a model or analogy that might be applied to the development of African thought.

Non-philosophy, for Okere, "must stand for the non-reflected, that unreflected baggage of cultural background" (1983, p. 88). "[F]or any reflection to take place at all, some data have to be taken initially for granted, without proof, and as it were by decree" (p. 88). Reflective reason, then, must have some unreflective material with which to work. This unreflective material is not simply a function of culture (in other words, it is not that so-called "traditional" cultures are necessarily the source of such non-philosophy), but non-philosophy is a feature of human finitude and the impossibility of "total rationalism," that is, the possibility that we could subject everything to reason.

Okere begins by outlining non-philosophy in Hegel. He sees three "non-philosophical presuppositions" in Hegel's philosophy of history (pp. 96–97). First, there is the assumption that the highest principle is Spirit or reason or idea. Second, there is the assumption that the world of nature is subordinate to the world of spirit. Third, there is the assumption of the absolute rule of reason in history and the dismissal of chance. All these, Okere argues, are taken from Christianity, and as such are non-philosophy since they come from faith rather than reason. It is also worth noting that Okere seems to equate "non-philosophical" and "irrational" (e.g., p. 97 and elsewhere), although we might also see the non-philosophical as related to the unreflective, the cultural, or the transcendent (Kanu, 2015, p. 20).

The function of non-philosophy for Okere and those who write about him is to provide the "stuff" that philosophy can interpret. "Religion and politics, but especially religion offered the non-philosophical material for Hegel's philosophical reflection" (1983, p. 99). He argues for a similar Christian grounding in Heidegger. Plato, of course, does not have a Christian non-philosophical basis, but is dependent on Greek religious thought. In each case, something must precede philosophy to provide the basis for universal reasoning. Okere is undoubtedly correct that each of these figures, and many others we might identify, are indebted to their cultural backgrounds, including their religious backgrounds. There are, though, some obvious questions we might raise about all this. For example, there could be, and most likely were, many other formative features for any of these philosophers. Not only that, but some of the formative features may well have been things that they did not even realise influenced them.

We might also raise questions about what kind of account Okere is giving us here.

- Is this an account of the creative process of the philosopher?
- Is the point that the philosopher must identify his or her own non-philosophies, and out of those create a philosophy?
- Is the point that we can, in hindsight, place the philosopher in a narrative?
- Is the point that interpreters of the philosopher are not doing their job unless they connect those non-philosophies with the resultant philosophies?
- Given that there are cultural and religious foundations for philosophies, are we saying that we can use those to adjudicate the adequacy of any philosopher's efforts?

On first glance, it looks as if what counts as non-philosophy for Okere is religious or cultural content. Philosophy becomes a way of processing this content, either in the sense of refining it to find the philosophical underpinnings or implications already implicit, or in the sense of providing the occasion for philosophy to engage in its activities. In Hegel's case, non-philosophy is the "material for Hegel's philosophical reflection" (p. 99). We might, then, ask more specifically about what it is in non-philosophy that makes philosophy possible. If it is the "material," does that mean that philosophy does not have its own material, and it needs something else? Will it never have its own material, or is it just that starting the whole process off in the first place requires something non-philosophical (this seems to be the picture we get from Okere about Heidegger – he would not have started without his roots in religion). And what extra value does philosophy give to these non-philosophical things – self-reflectiveness? Logical rigour? Universality? Are we sure that these things did not exist prior to philosophy's engagement with them, in the cultural and religious material itself? In other words, does philosophy create the rigour or universality, or does it merely find it?

And, what of philosophies by others which do not clearly have a religious or cultural basis? If we haven't yet found a connection to religion or culture, does that just mean we haven't looked hard enough? The thinkers Okere has chosen tend to take us back to not only religious roots, but Christian roots, including concepts such as salvation, faith, and so forth. Must philosophy have these things, if its engagement is with non-philosophy, or is that an accident of history?

And do we necessarily have an archaeological direction here? In other words, is it the case that philosophy looks to its intellectual (if not historical) past to find its foundations? And if there is archaeology, is there also teleology, that is, is there a necessary direction forward for philosophy? Certainly at times in the past philosophy has conceived of itself in these terms, but is that still a credible account today? In other words, if the impetus for philosophy was to fulfil some aspect of non-philosophy, in the form of religious concepts such as salvation (as Okere suggests when he outlines Hegel's path), does philosophy retain that agenda, and should we be looking for philosophy to save us?

It is worth also thinking about what happens once we have identified non-philosophy. In Okere's model, all philosophy starts from cultural and religious content. But what does philosophy do with this content? Okere's *African Philosophy* gives his answer for this (a hermeneutical analysis), but it is worth noting that one could equally give an analytic philosophical method which distilled philosophy out of non-philosophy. His first chapter, in which he surveys some "false routes" in African philosophy (including Tempels, Kagame, and Mbiti), concludes that early efforts to do African philosophy failed because they did not answer some basic questions, specifically about the nature of philosophy and the difference between philosophy and non-philosophy. The result was that these authors were unable to start the philosophical task because they were confusing culture with philosophy.

The reason for raising these questions is to be more specific about what is interesting in Okere's proposal. Non-philosophy, for him, is the reason philosophy can do what it does. He might be right about this. But the case he makes for it, using the specifics of some European philosophers in a specific tradition, suggesting that there is mainly a religious basis for philosophy – that seems less likely. So if we can extract the central question about the nature of non-philosophy and its relationship with philosophy from his specific account of why it matters, we might make some headway.

Alternatives to Okere's model

There is one more stop to make before we get back to thinking directly about womanist and environmental philosophy, another alternative to the kinds of non-philosophy sketched here and, in particular, to Okere's version. Okere is not wrong in his depiction of non-philosophy, but the specific way he frames it artificially limits it. In so doing, I believe, he makes it difficult for African philosophers to think about womanist and environmental philosophy, along with other areas that have not yet received much attention. In Okere's model, I believe we are faced with the expanding circle model of philosophical engagement with new areas.

I would like to suggest an alternate way of thinking about non-philosophy that will take us in a different direction. Rather than seeing non-philosophy as the "stuff" that philosophy uses to produce its rational accounts of the world, what if we saw non-philosophy as the ever-present interrogative moment on philosophy itself? What if it lies not just at the beginning of the philosophical process, but at every point? I have argued elsewhere that philosophy is a mode of questioning (a claim which itself might be open to question, which would only validate it) (Janz, 2009a, pp. 217ff). That questioning is sometimes seen as dialectical, as the moment that starts a line of inquiry and which is ultimately resolved or transcended. The questioning I have in mind does not end, but remains throughout the philosophical process of striving for universals and for coherent accounts of the world. The expanding circle model of philosophical progress is the model of extending the rationality of philosophy to new areas, and at the same time rectifying philosophy's past blind spots in its own logical structures. Philosophers made claims about rationality for humans, and yet left out women, non-Europeans, and others. The expanding circle rectifies that error and strengthens philosophy's claim to reason. Ethical agency and standing had been accorded to humans, and then we realise that there might be a case in extending that to non-humans – animals, ecosystems, the environment in general. The expanding circle again rectifies philosophy's claims to reason.

What we do not have here is the preservation of non-philosophy. Instead, it is seen as a defect. The questioning of philosophy itself leads to its own rectification, and thus the end of the question. My interest is in the questions that remain within philosophy, not because they are difficult or because we haven't seen them yet or been able to answer them, but because every answer we give only closes off other answers, settles, for the moment, the question but does not answer it in the sense of stopping it. This is the non-philosophy that is more interesting, the kind that continues. It can be seen when philosophy's own assumptions are questioned by the very things that it has tried to include in the expanding circle.

Okere gives us a kind of questioning, rooted in the Heideggerian *Seinsfrage*, which is intended as reflexive and rooted in his version of non-philosophy, that is, the raw material found in culture and religion. But the *Seinsfrage* is not the questioning I have in mind here. I am not talking about dialogue. This form of questioning is predicated on human meaning, as is the entire hermeneutic tradition. To inquire about the environment in this manner means to inquire about the human environment, the environmental space that has some implication for humans. It is the *Umwelt*, the space of Dasein, Gadamer's encounter at the horizon. It might be Other (although it might not), but it is not apart. Philosophy in this mode remains human centred.

This is not to say that hermeneutics is not useful or cannot give insight, but just that it leaves out a part of the philosophical production of concepts that most effectively addresses non-philosophy. The hermeneutic tendency is to turn non-philosophy into philosophy, and by so doing it loses its potency. Culture, religion, sensation, experience, whatever we call non-philosophy, is tamed by being understood, in the sense that it becomes represented. We see this in part if we think about hermeneutic's own view of itself. It is presented as universal, or rather universally applicable, but rarely does it consider the questions it answers, the culture from which it comes. In the West, hermeneutics arises in tandem with the rise of modernity, with questions about the place of technology and science in relation to human experience and meaning. It is an answer to a specific threat, a specific non-philosophy. In answering as it does, it privileges some kinds of texts, some kinds of philosophical pursuits, and ignores others. Does hermeneutics in Africa have to answer the same questions, in its constitution (as opposed to in its use)? Not necessarily. Its own non-philosophy might be entirely different. Okere does not take up that question, but it is a significant one, since a philosophical approach is not beyond questioning.

In philosophy, everything is potentially and theoretically able to be the object of philosophical reflection, but not everything is. From our vantage point today, we rightly decry the oversights and political moves of the past which served to marginalise some areas that we now consider to be important. And yet, this simply remains a contingent artefact of our time and context. Decades from now, later philosophers will no doubt decry our lack of insight in overlooking something else.

This does not mean that since there is contingency here, there is arbitrariness and it does not really matter what we pursue. It does matter. But it also matters that we move away from the expanding circle of determining the areas worthy of philosophical attention. The expanding circle does not, in fact, expand; it simply moves. There is always peripherality whenever there is a rational gaze. The reasons for including womanist philosophy and environmental philosophy are not simply that they have been left out in the past and that we want everything included in a grand conversation. It is not because those who have received disciplinary recognition in African philosophy now feel bad and want to make up for previous exclusions.

We might also de-centre Okere's version of non-philosophy, grounded in culture and religion, and ask what else might be the occasion for thought. After all, both culture and religion are already meaningful things in a specific sort of way. They often have narratives, for instance. They often explicitly involve values. They are, in other words, tailor-made for philosophising. But what about things that do not seem to come with these ready-made connections? What about objects, for instance? The materiality of objects might be given narrative, but we might ask about the meaning, nature, and status of those objects before those narratives. It might be difficult to see that any one object impinges much on our philosophical practice, but we do know that the cumulative effect of objects on us constructs our spaces, opens some vistas for thought and closes down others, and generally orients us toward particular kinds of narratives and away from others. Those objects come into being through the action of humans, or of animals, or of natural processes, and in each case they are the trace of forms of knowledge which might not have risen to the level of conscious or rational reflective thought.

We might, in other words, imagine an African object-oriented philosophy, perhaps not in the sense that it has developed in Europe, but in its own sense. We can think about the difference between the useful and the useless, the desired and undesirable, the comfortable and the unsettling. We can think about difference, that is, how objects both support our forms of thought and also push us into revising and inventing new forms. We can think about what constitutes an object – not a type of object, but an object – and what virtualities or events it might bring into being by its presence. All this surely raises questions about gender and about the environment within an African context without resorting to the expanding circle, since those affects and forms of engagement we have, which support our thought-lives, are experienced differently by different people. Questions of culture and religion remain important, but they are not non-philosophy any more. They are now alternate modes of thought, alongside philosophy, which are not overcome as philosophy trains its rational gaze on the world. This example of objects is just that – an example. There are others. The point here is to think about the ways in which non-philosophy remains within all philosophising, and therefore how areas of inquiry such as women's experience and the environment are not simply problems to be solved, but engines of new thought within philosophy, as they continue to exercise an element of non-philosophy at all times.

Womanist philosophy and environmental philosophy

The choices of African philosophy, as is the case for the choices of other places of philosophy, have been as much about politics as about philosophy. If, as I have long argued, the central question of African philosophy has for too long been "Is there an African philosophy?" then those who explicitly or implicitly take that question seriously will choose their pursuits in philosophy in order to answer that question. This is at least one reason why we have not seen much focus on womanist philosophy and environmental philosophy – those pursuits would not help to answer that question, at least in the minds of those who take the question seriously. Carving a space on the map of world philosophy was a political task (and, it should be said, the rest of the world of philosophy has been no less political, even if unconsciously so). In a sense, then, the choice of how to pursue philosophy had an element of non-philosophy within it – the imperative of establishing African philosophy's right to be taken seriously. This imperative was and is, of course, understandable and defensible, but it is still non-philosophical.

In the title of this chapter, I refer to "provocations" such as womanist and environmental philosophy. This is a way of signalling a kind of resistance to the expanding circle approach and instead recognising the ways in which peripherality and non-philosophy are significant. It is important to recognise that I am not claiming that these areas are peripheral to African philosophy, nor that they are the non-philosophy to the "real" philosophy that is the mainstream. The point rather is that these areas stand in a unique position, one of enabling the possibility of new questions. These new questions are not simply ones of inclusion – "what about women?"; "what implications does this have for the environment?" – those questions would be little better than the misguided question "Is there an African philosophy?" Neither are they questions of application, as if we can work out our metaphysical, ethics, or politics first and then just apply them to these areas.

We exist in multiple places at once, both in terms of the level of abstraction we use in thinking about place and in the overlapping and intersecting forms of scaffolding which structures our thought. When we take the expanding circle model of philosophical inclusion, we work at a long focal length. We think about "women" as an abstract category, and ask how that category can be included in philosophy. The same is true of the environment. We start, in other words, like classic philosophers – assuming that our abstractions designate categories that have a kind of integrity, and which we can work out at the abstract level.

Doing this, though, simply replicates (or at least threatens to replicate) the androcentric, anthropocentric thought we have always had. We do not question our practices or assumptions; we simply ask how they can be modified so as to include a previously ignored group or thing. This does not suggest that all assumptions must be questioned at all times, but, on the other hand, we should also not assume that a practice of philosophy is immune to those questions. Philosophy, more than any other pursuit, always returns to its beginnings; however those are understood. And, to not inquire about philosophy's assumptions comes at a cost. At its worst, the assumptions which are reinforced can be racist, misogynistic, nativist, and so forth. "Ecology" as a term was, after all, coined by Ernst Haeckel in the 19th century and linked a concern for nature with a racist belief in whose nature it was. It was the foundation of the Nazi *Blut und Boden* ideology during the Second World War, a kind of ecofascist belief that the purity of nature must be preserved and in doing so it would elevate the purity of the Aryan spirit. He introduced a particular approach to Darwin into the German-speaking world, one which held that human evolution happened in independent and parallel tracks we now know as races. All of this was the result of a kind of environmental thought that was never allowed to change the direction of established concepts. The same is at least in principle possible when including women in philosophy. It is possible to imagine a situation in which women are simply included, but the fundamental nature of philosophy is never questioned.

But what is the alternative? It is, I believe, in a particular form of listening. Instead of asking about a long focal length category like "women," we could just listen to actual women. But what does that mean? Have we not been doing that? Perhaps, in some sense, but listening should be as much a reflexive activity as an outward-looking one. In other words, listening means in part questioning ourselves, our own conceptual assumptions. It means taking seriously the cognitive scaffolding that sustains others. Listening has been comparatively understudied, despite there being several books and an *International Journal of Listening* available. To listen means in part to allow for silence, especially one's own. Listening does not just mean listening to someone else's point of view, but also listening to someone else's way of engaging the objects of the world.

Does one listen to the environment? I have argued elsewhere that we do (Janz, 2009b). Having a knowledge and an ethics of place requires that we recognise the knowledge systems that are not our own, the traces that have been encoded by non-human action in the world. It means looking closely, and the more closely we look, the more we are likely to question our own abstractions about nature and our relationship to it. We have already started to question our assumption about dominance over nature, given to us by the enlightenment. That is an excellent first step, but there is more to be done.

Listening is not, I think, just the hope of understanding the other. It is, rather, being moved by the other. "Being moved" does not necessarily mean being emotionally moved, although it might. It means having our trajectory changed. This is a different impulse from the "expanding circle" idea – accommodation isn't the same as changing our trajectory. We are moved when the conceptual ecology in which we work has to take something new into account. It is revision and adaptation, rather than expansion and accommodation. If there is little history of women's presence in philosophy, the point is not to just extend philosophical methods and concepts to "women's issues," as if these are a separate and unique species of non-philosophy meant to be tamed by philosophical method, but rather to think about what difference it makes in asking philosophy's questions to take seriously women's experience. Are there forms of writing, forms of materiality and practice, ways of knowing, that have been accorded secondary status in philosophy in the past, but which are more adequate to expressing or capturing women's experience? Then philosophy's way of asking its own questions must be shifted to take into account a different mode of expression. Are there forms of materiality in nature which encode knowledge in ways that humans are not familiar with? What would it mean to apprehend those in their own terms, rather than for our purposes? Can a tree know? Does an ecosystem have memory? What would it mean to de-centre human interests in our pursuit of all philosophy? What happens when we take peripherality seriously, that is, the milieu that sustains our thought? What new questions might be created if we are moved in this way by nature?

What is Africa's own non-philosophy? I don't think it is limited to the cultural or religious grounding to which Okere pointed. There is no archaeology or teleology. It is not the promise of practicality or applicability, the "relevance" that others hope for. It is the ongoing intellectual challenge from other disciplines and from lived experience in Africa by those committed to this place. It is the set of concepts created by those disciplines and those people, scaffolded in various ways by their habits and practices, institutions, language, and bodily engagement with the world. Philosophers do not start with abstractions and universals; we earn them. We create them. We do that not by expanding circles, but by being moved in new ways, if we are open to it. It is the bravery of philosophers to ask difficult questions and to allow settled forms of thought to be challenged. It is the humility of philosophers to not spare ourselves from any of this.

References

Gallagher, S., Reinerman-Jones, L., Janz, B., Bockelman, P., and Trempler, J. (2015). *A Neurophenomenology of Awe and Wonder: Towards a Non-Reductionist Cognitive Science*. Houndmills, Basingstoke, Hamshire: Palgrave Macmillan.

Janz, B. B. (2009a). *Philosophy in an African Place*. Lanham MD: Lexington Books.

Janz, B. B. (2009b). Thinking Like a Mountain: Ethics and Place as Travelling Concepts. In M. Drenthen, J. Keulartz and J. Proctor (Eds.), *New Visions of Nature: Complexity and Authenticity*. Series: The International Library of Environmental, Agricultural and Food Ethics (pp. 181–195). New York: Springer.

Kanu, I. A. (2015, June). African Philosophy as an Ontologico-Existential Hermeneutics. *Igwebuike: An African Journal of Art and Humanities*, 1(1), 17–24.

Laruelle, F. (1999). A Summary of Non-Philosophy. *Pli*, 8, 138–148.

Laruelle, F. (2010). *Philosophies of Difference: A Critical Introduction to Non-Philosophy*. London: Continuum Publishing.

Maoilearca, J. Ó. (2015). *All Thoughts Are Equal: Laruelle and Nonhuman Philosophy*. Minneapolis, MN: University of Minnesota Press.

Okere, T. (1983). *African Philosophy: A Historico-Hermeneutical Investigation of the Condition of Its Possibility*. Lanham, MD: University Press of America.

Park, P. K. J. (2013). *Africa, Asia, and the History of Philosophy: Racism in the Formation of the Philosophical Canon, 1780–1830*. Albany: SUNY Press.

Ramose, M. (2015). On the Contested Meaning of 'Philosophy'. *South African Journal of Philosophy*, 34(4), 551–558.

2 Gaia hypothesis from an African perspective

Alexander O. E. Animalu and Jeff Unaegbu

Introduction

We anchor the complementary duality of African (Igbo) three-valued thought logic on the Igbo-Ukwu bronze torus artefact (c. 1,000–5, 000 BC) representation of a solution of the well-known three-body problem of classical mechanics, namely the problem of the system of the Sun (Father-Nna), Earth (Mother-Nne) and Moon (Son-Nwa) and discuss the implications of the 3N "linguistic geometry" for the marginalisation of women and the environment in African philosophy, in which the sad fate in Western (Greek) culture of the Earth goddess (Gaia) and her python symbol, along with her African (Igbo) roots evident in the cognate Igbo names, relates to her futile attempt to invent medicine from the Earth's environment to make humans immortal.

Gaia is the name of the Greek mother goddess of the Primeval Sea. The Gaia hypothesis in science, due to Lovelock (1979), is also known as the Gaia theory or Gaia principle and proposes that all organisms and their inorganic surroundings on Earth are closely integrated to form a single and self-regulating complex system, maintaining the conditions for life on the planet. The mythical Gaia is a female or mother which inspires the idea of Gaia hypothesis that looks at the earth environment as a mother of some sort producing and supporting all organisms. The lead question is usually whether human activities are capable of affecting or reducing the viability of this system? This falls within the domain of environmental philosophy. James Lovelock and Sidney Epton (1994) not only believe this is the case, a fortiori, they suggest that lack of adequate attention in scholarship to the danger which the environment faces is worrisome. Like the neglect which the mythical Gaia suffers, we ask, is there a correlation between the fate of marginalisation suffered by the mythical Greek female, Gaia and the Igbo-African mythical female ichie? Does this unfortunate fate have to do with the fact that they were both females? Represented as a female or mother in both the Greek and the Igbo thought, does the theme of environment suffer marginalisation in philosophy because of this feminine connection? Our objective in this work would be to show that the marginalisation of the environment in philosophy generally, and African philosophy specifically, may be a direct consequence of its perception as a female, the subordinated sex. The hypothesis has analogies and various interpretations in social sciences, politics, and religion under its dual two-value thought/logic (Daisyworld positive/negative feedback model of Gaia theory) (Watson and Lovelock, 1983) and philosophy. For this reason, there are conceptual and linguistic difficulties in naming a new science that unifies living and non-living parts of the planet Earth after a Greek goddess, Gaia (Gaea or Gaya), which we shall discuss in Sec. 2 in order to identify the sad fate of the goddess in Greek mythology and its linguistic implication in the language and geometrical foundation of Greek philosophy in the marginalisation of women. In Sec. 3 we shall use the complementary duality of African three-valued thought/logic and philosophy (Animalu and Chimakonam, 2012) anchored on a representation of the three-body problem of the system of the Sun, Earth, and Moon to characterise the marginalisation of women in emerging African Philosophy. In Sec. 3, we shall use "linguistic geometry" to discuss the marginalisation of women in both Greek and African philosophy vis-à-vis the attempts from 19th century by Western European historians of philosophy to deny, as pointed out by Onyewuenyi (1993) on pages x and 103 of his book, "that the Greeks learned philosophy along with geometry and other sciences from the Egyptians [i.e., Africans]." The chapter will be discussed and concluded in Sec. 4.

Greek Gaia mythology and Gaia science

In order to relate the Greek myth concerning Gaia to the Gaia hypothesis, we begin by exhibiting, firstly, the African (Egyptian) roots of Greek geometry as elaborated in Figure 2.1 based on the challenge in the 19th-century history of philosophy when Africa's relationship to the Western world was dictated by racial and romantic feelings on "how to measure the heights of the pyramids by their shadows" on page 105 of Onyewuenyi (1993). Note the duality in the elaborated pentagon shadow of the square pyramid defined by the five corner points of its square base and the tip.

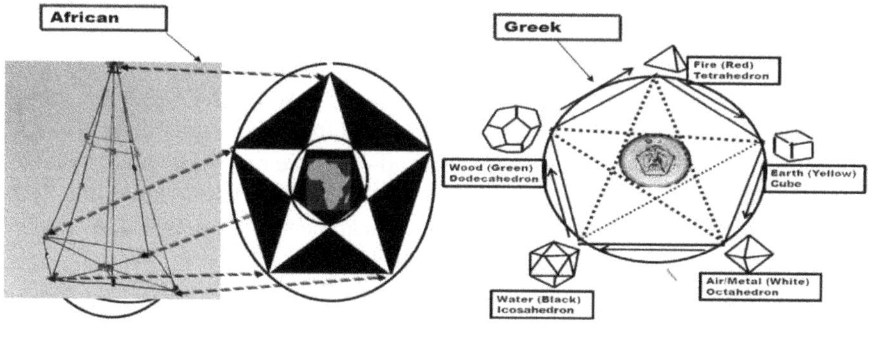

Figure 2.1 Pentagon shadow of the (African) square pyramid geometry (c. 1100 BC) used by the Greeks (c. 300 BC) in the chart of elements with Platonic geometries in Plato's model of the earth, and later by Kepler (c. 1600 AD) in his cosmic pentagon grid over a spherical solar system

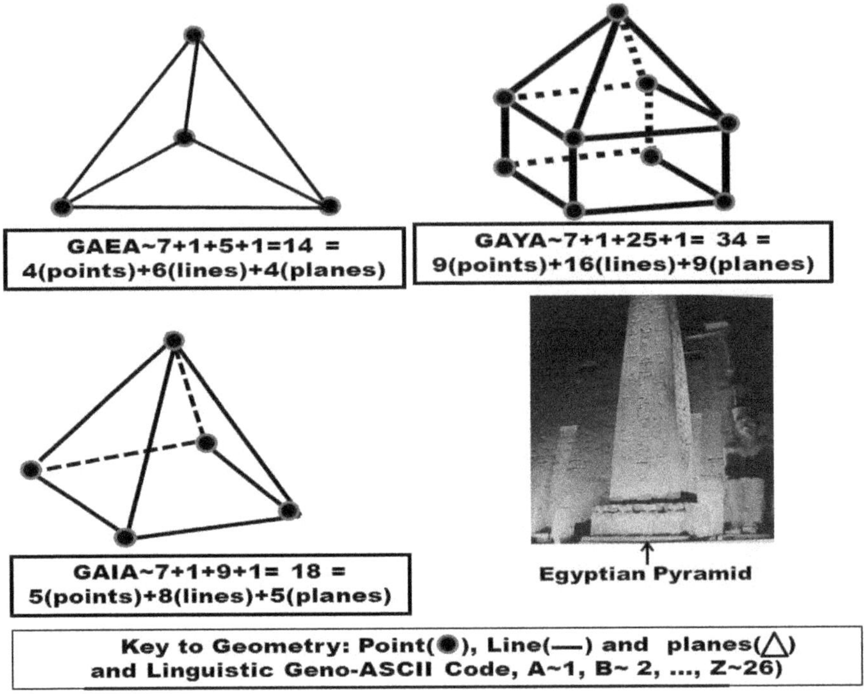

Figure 2.2 "Linguistic Geometry" of GAEA, GAIA & GAYA in relation to the Egyptian pyramids

Secondly, we show in Figure 2.2 the relationship to the pyramid geometry of the names, Gaia, Gaea and Gaya in "linguistic geometry" where GENO-ASCII Code (developed by A.O.E. Animalu and A.V. Animalu (2012) for the translation of Alphabets into natural number equivalents and geometry with A~1, B~2, C~3, . . ., Z~26, and WORD~W + O + R + D = 23 + 15 + 18 + 4 = 60) gives GAEA~7 + 1 + 5 + 1 = 14, GAIA~7 + 1 + 9 + 1 = 18 and GAYA~7 + 1 + 25 + 1 = 34 i.e., progressive generalisations of a pyramid from a triangular base (GAEA) to a square base (GAIA) to a square-pyramid-on-cubic box (GAYA). It is therefore reasonable to capture this geometrical "evolution" under the name Gaia hypothesis (or Pyramid principle, from an African perspective) as well as to note the duality between the number of points and planes in the ratios,

points:lines:planes = 4:6:4 (GAEA), 5:8:5 (GAIA) & 9:16:9 (GAYA)

while lines are self-dual in accordance with the principles of projective geometry.

It is also pertinent to point out that, to Pythagoras, who lived and studied in Egypt and whose knowledge of geometry and medicine is said to be of Egyptian origin, and to his followers known as the Pythagoreans (500–600 BC) (Farrington,

1953), "the function of geometry is to draw us away from the world of the senses, to a world of the intellect and the eternal. For the contemplation of the eternal is the end of philosophy as the contemplation of the mysteries (of creation) is the end of religion."

According to the ancient Greek mythology at the Archaeological Museum Site at Kos, Greece, Gaya was a goddess of the Earth and once human. When she was human, she loved her fellow humans so much that she taught them how to use materials from Earth's environment as medicine and the art of healing. In a bid to learn and teach humans immortality, she was exposed to the ire of other gods and goddesses who created circumstances in which she was defiled by a mortal lover who subsequently died, and in her attempt to revive the lover with her snake wound round a walking cane that healed by touch and a cup full of medicine, she was beheaded by another goddess. The significance of this myth is evident in the sculpture of beheaded Gaya at the Archaeological Museum Site at Kos, and her son, Asklepios, carrying her walking cane with healing snake wound round it. For lack of space, we cannot reproduce this figure here.

The authenticity is evident on a visit to the Archaeological Museum Site where there is the image of Gaya without the head, which an inquisitive tourist can see lying beneath the sculpture but covered so as to discourage photographing. Gaya's art house and its dating (170–160 BC) is also in the neighbourhood of the beheaded sculpture of Gaya. The punishment of the female goddess in this myth and its echo as marginalisation of women in Greek philosophy has a parallel in the African (Igbo) mythology, thought and philosophy to which we now turn.

African myth, thought logic and philosophy

In order to relate the implication of the Gaia hypothesis for the marginalisation of women and the environment in African philosophy, we now proceed to apply the proceeding "linguistic geometry" to the three-valued (3N linguistic) thought logic (Nne, Nna, Nwa) anchored (Animalu and Chimakonam, 2012) on (Igbo) cosmic consciousness of Earth~Ani, Sun~Anyanwu and Moon~Onwa and African philosophy preserved in the artefacts of "ichi" and bronze torus. Again, for lack of space, we cannot reproduce this figure here.

Otherwise, it would have been easy for us to see how the facial scarification ("ichi") as the "face of the sun" of the Igbo-Ukwu craftsman and his cosmic consciousness are brought to bear on the sculpture preserved in the bronze torus artifact (c. 1,000–5, 000 BC) representing, so to speak, a geometrical solution of the well-known three-body problem of the Sun, Earth and Moon in classical mechanics and the complementary duality with respect to the interchange of the Sun and the Earth in two-dimensional projection (like shadow of the pyramid in Figure 2.1).

Finally, by using the GENO-ASCII Code, we observe that the two names GAYA~7 + 1 + 25 + 1 = 34 and ICHIE~9 + 3 + 8 + 9 + 5 = 34 are numerically cognate. Although there are instances of numerical cognate words, which like synonyms, refer to different things, such a "cherry-picking" is based here on geometry rather than on pronunciation (sound) and on the fact that just as "Ichie" refers to an African(Igbo) myth of a divine feminine (Animalu and Acholonu,

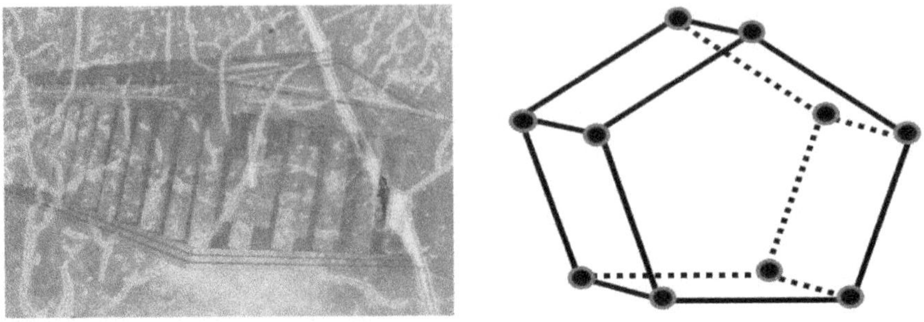

Figure 2.3 Gaya's art house (Temple of Asklepios) (170–160 BC) and its geometry characterised by the number of points: lines: planes = 10 + 18 + 7 = 32 represents LIFE~12 + 9 + 6 + 5 = 32 in GENO-ASCII CODE

2009) so does the Gaya goddess in the Greek myth. The myth of the divine feminine ("Ichie") is characterised by the Idemmili ("pillar of water") goddess with facial scarification, whose messenger is the python (Eke Idemmili) described by the novelist Achebe (1969). From this perspective, one could identify the missing head of Gaya with the scarified head of the divine feminine ("Ichie") to arrive at the dual unification of the sculptures of the Greek Gaya and the African Ichie.

The common thread in the unified sculpture is the similarity of the life styles of Gaya and Ichie and their healing (snake) symbols, which led to their marginalisation through death as humans and other consequences, like general discrimination by male humans against female leadership.

Discussion and conclusion

We have shown that there is a correspondence between the Gaia hypothesis or theory in planetary science, named after a Greek goddess (Gaia), and the broader Pyramid Principle (in the African perspective) derived from the "linguistic geometry" of GAEA, GAIA & GAYA with respect to the Egyptian pyramid and its other African three-level thought logic and philosophy represented by the (Igbo-Ukwu) bronze torus. We have also shown that there is a corresponding marginalisation (as a euphemism for death) in Greek mythology of the goddess (Gaya) and the African (Igbo) counterpart of the divine feminine ("Ichie"), in the course of their attempts to invent medicine from the Earth's environment to make humans immortal. The fact that the son Asklepios suffered the same fate as the mother means that marginalisation as a euphemism for death applies to all mortals: this gives meaning of LIFE to Gaya's art house (temple of Asklepios) (see, Fig. 2.3).

We conclude that the marginalisation of women (as a euphemism for death) is also applicable to men, as a law of nature against immortality of the living, but the circumstances are dependent on the environment of life.Thus as Gaia in the Greek mythology inspires a theory of the environment in which the earth environment is perceived as a mother that gave birth to, and supports all other things in nature, the ichie or nne in the Igbo-African mythology standing for the earth goddess is thought to be the mother of all things in nature by the Igbo. Coincidentally, the mistreatment of the Greek Gaia which culminates in her beheading correlates with the mistreatment of the Igbo ichie which culminates in her facial scarification. These two myths are symbolic representations of the discriminatory tendencies in both Greek/Western and Igbo/African cultures against women. As the earth environment is depicted with the attributes of the female sex in both myths and in Gaia hypothesis, so also does the marginalisation of women in both cultures translate to the marginalisation of the environment in both the West and Africa, in philosophy and in African philosophy specifically. Our conclusion is that the challenges we face in the battle to save the environmental in the sub-Saharan Africa today are in part, negated by some hard line cultural beliefs and practices. A change of attitude is therefore recommended.

References

Achebe, C. (1969). *Arrow of God* (p. 81). Dobleday and Anchor, NY, Revised Version with picture of the python as coverpage.

Animalu, A. O. E., and Acholonu, C. O. (2009). 'Ichi' Linguistic Geometry and Evolution. *African Journal of Physics*, 2, 373–392.

Animalu, A. O. E., and Animalu, A. V. (2012). GENO-ASCII Code for Translation of Alphabets Into Natural Number Equivalents and Geometry. *African Journal of Physics*, 5, 122–140.

Animalu, A. O. E., and Chimakonam, J. O. (2012). 4x4 Magic Square Representation of Complementary Duality of African Three-Valued Thought Logic. *African Journal of Physics*, 5, 141–158.

Farrington, B. (1953). *Greek Science*. London: Penguin Books.

Lovelock, J. E. (1979). *Gaia: A New Look at Life on Earth*. Oxford: Oxford University Press.

Lovelock, J. E. and Epton, S. (1994). In Quest for Gaia. In L. P. Pojman (Ed.), Environmental Ethics: Readings in Theory and Application. Boston: Jones and Bartlett Publishers.

Onyewuenyi, I. (1993). *The African Origin of Greek Philosophy: An Exercise in Afrocentricism* (p. x and p. 103). Nsukka: University of Nigeria Press (quoting Sir Isaac Newton's *Principia*).

Watson, A. J., and Lovelock, J. E. (1983). Biological Homeostasis of the Global Environment: The Parable of Daisyworld. *Tellus*, 35B, 284–289.

3 How to ground animal rights on African values

A constructive approach[1]

Thaddeus Metz

Introduction

Kai Horsthemke's *Animals and African Ethics* (2015) currently stands as the most thorough reflection on the implications of characteristically African values for animal rights, where by 'animal rights' I mean a broad view according to which we have direct duties not to seriously harm animals, at least not for purposes such as enjoying the taste of meat, liking the look of their skins worn as adornment, and ensuring that cosmetics are safe for human beings. African values have often been deemed to be unable to justify animal rights, so construed (e.g., Callicott, 1994, p. 158), and, while there have been recent attempts to argue that they can, in his book Horsthemke systematically argues that they cannot.

I am one of those who has contended that certain values salient in the African tradition promise to ground animal rights (see Metz, 2012 and elsewhere in this volume; work by others includes LenkaBula, 2008; Behrens, 2010, 2014; Molefe, 2015). Drawing particularly on recurrent sub-Saharan themes, I have advanced the claim that, roughly, a being has a greater moral status the more it is by its nature capable of being part of a communal relationship with characteristic human beings, and I have invoked this conception of moral status to derive a number of direct duties towards animals. In *Animals and African Ethics* Horsthemke advances several objections to my approach, which he calls 'African Modal Relationalism,' arguing that it is anthropocentric and speciesist, is poorly motivated relative to his egalitarian-individualist approach, and does not have the implications that I contend (2015, esp. pp. 85–92, 99, 123, 144). My aim in this chapter is to rebut these criticisms.

I begin by providing a sympathetic overview of Horsthemke's book, which I contend merits engagement, despite our disagreements about the plausibility of my views (section 2). Next, I sketch the core of African Modal Relationalism, presenting enough of it to facilitate debate between Horsthemke and myself (section 3). I then spell out Horsthemke's objections to this theory and reply to them (section 4). I conclude by pointing to the nub of many of our disputes and what would need to be done in future work to take the debate still further (section 5).

Overview of *Animals and African Ethics*

Animals and African Ethics is the first book-length treatment of salient African moral perspectives as they bear specifically on animals. By values (and norms) that are 'African,' I and Horsthemke (2015, pp. 2–3) both mean ones that have been salient below the Sahara Desert, or otherwise indigenous to Africa (and so not a function of Christianity or Islam), amongst a wide array of peoples and over a long span of time. Neither of us is committed to 'essentialism,' the view that certain values always inhere in something African, or only inhere in something African. Instead, we maintain that there are certain conceptions of the good and the right that have been recurrent amongst traditional African peoples in a way they have tended not to be amongst others in the world. While other scholars have explored environmentalism in general from the perspective of African values, so construed (e.g., Behrens, 2011; Kelbessa, 2011; Chemhuru, 2016), Horsthemke is the first to have focused on the moral status and proper treatment of animals in particular.

In his book Horsthemke provides careful, well-researched analyses not merely of traditional African values, but also of how they plausibly bear on contemporary moral and legal debates about animals, often ones that are prominent on the continent. For instance, he takes up issues such as the slaughter of animals for ritual purposes (say, to pay respect to ancestors), the eating of bush meat, and the use of what in southern Africa is often called '*muti*,' animal parts believed to have the powers either to ward off harm or to cure disease.

With regard to these and similar issues, Horsthemke maintains that such practices are typically immoral and that African values cannot make sense of such a judgement. In reaching this conclusion, Horsthemke does not dismiss the African tradition of thought about animals, but rather takes the time to critically engage with a variety of its recurrent and important manifestations. He considers, and rejects, views commonly held by indigenous black peoples, contending that just as contemporary thought in Africa rightly tends to abjure the sexism recurrent in traditional African ways of life, so it should likewise forego the speciesism that Horsthemke finds in it. In addition, Horsthemke addresses philosophies with an African pedigree that have recently been articulated by academics, finding them to be similarly speciesist or anthropocentric, on the one hand, or too vague to apply, on the other.

By virtue of Horsthemke's in-depth, thoughtful, and contentious reasoning, any future Africa-inspired thought about duties towards animals must engage with his book. For theorists like me, who both work within the African philosophical tradition and believe in animal rights, his book presents a challenge; for it argues that we cannot coherently do both.

African Modal Relationalism

In developing what Horsthemke calls 'African Modal Relationalism,' one of my intentions was to probe particularly sub-Saharan worldviews for ideas that would provide a plausible ground for animal rights. I wanted to advance a theory that would be useful to someone, supposing she wanted a way to invoke indigenous African values to make sense of direct duties not to hunt, not to factory farm, not to use animals for trivial experiments, not to kill animals for the sake of clothing, etc. In this section, I recount three ideas about morality that have been salient in the African tradition, and then indicate how I have drawn on them to create a prima facie attractive philosophy of animal rights.

One idea recurrent amongst indigenous African peoples has been that entities in the world are ordered in a graded way or form a 'chain of being' (Olupona, 1991; Teffo and Roux, 1998, pp. 196, 198–201; Chitando, Adogame and Bateye, 2012). At the top is God, the source of everything else in the world, after which come ancestors, wise founders of a clan who have survived the deaths of their bodies and continue to reside on earth. Next come human beings, and then animals, plants and rocks, in descending order.

Sometimes this gradation is conceived metaphysically, in terms of vital force, an invisible energy thought to have its source in God and to permeate everything that exists in various degrees. Other times, however, it is conceived ethically, as a ranking of final value or moral status. Traditionally speaking, the ethical is usually grounded on the metaphysical, so that the more vital force a being has, the more importance it also has.

Such a ranking does imply that human beings have a greater moral importance than animals. However, the latter have their own, independent claim to ethical treatment, as do even elements of the vegetable and mineral kingdoms. Despite the hierarchy, it is easy for this worldview to forbid stomping on a caterpillar for the mere fun of it and more generally to prescribe a sense of reverence for all that exists.

A second common theme about ethics in traditional African thought is the idea that relationship is central to becoming a good person. Moral behaviour is ultimately, or at least mainly, a matter of creating, sustaining and enriching relationships (e.g., Teffo and Roux, 1998, p. 204; Christians, 2004, pp. 244–245; Murove, 2007; Metz, 2013; Metz and Miller, 2016). Such a view is reminiscent of the ethic of care, but, unlike the latter, which is a relative newcomer to the West and is a minority view there, a relational ethic has been prominent in African philosophy for at least several hundred years.

A third thought about how to live that is salient amongst pre-colonial sub-Saharans is communitarianism or communalism (for overviews, see Kasenene, 2000, pp. 349, 354–355; Odimegwu, 2007; Wiredu, 2008). Often African philosophers make claims such as that the community is 'prior to' the individual. Although that is sometimes understood to mean that the group has moral pre-eminence (corporatism), I prefer the reading according to which a person cannot develop human excellence except by entering into communion with others (relationalism).

In drawing on these ideas, I do not always take them on board as they have been traditionally stated. My project has involved not simply representing African thought and applying it, but reinterpreting it as necessary to create a philosophically powerful theory, one that would resonate with reflective people throughout the world, not merely those partial towards African worldviews. Much of this means that I abjure appeals to the existence of invisible agents or forces such as ancestors, witchcraft or vital force (or what Westerners would call the 'spiritual').

African Modal Relationalism is in the first instance an account of moral status, i.e., of what in the world is owed moral treatment for its own sake or, equivalently, what is the object of direct duties. According to this account, a being has a greater moral status the more it is by its nature capable of relating communally with characteristic human beings. I now spell out the key concepts involved, here.

Appealing to elements of the notion of a chain of being, this principle includes the idea of degrees of moral status. Some beings are morally more important than others. However, unlike the traditional conception, as will become clear below, it does not entail that everything in the world has some moral status.

Moral status is a function of the extent to which a being can relate to us in a certain way. Instead of appealing to a highly contested metaphysical notion of vital force, for which African philosophers have yet to provide much evidence, the principle considers moral importance to be determined by a secular concept of the capacity to enter into relationship. This approach is why Horsthemke's label of 'modal' is apt; it is not actually being a part of a relationship, but rather a being's *ability* in principle to be part of a relationship that counts.

Finally, according to the principle, the relevant relationship is a communal one. Instead of a corporatist interpretation of the African tradition, the principle entails that it is a particular entity, and not a group, that has a moral status. Specifically, a being's moral status is determined by the extent to which it is the kind of thing that can be party to a relationship in which there is communion, understood in terms of identifying with others and exhibiting solidarity with them. That is, a being's moral status is graded according to its ability to be part of a relationship in which there is a sense of togetherness and participation on a cooperative basis (identity), as well as aid consequent to other-regarding attitudes such as sympathy and altruism (solidarity).

There are differential degrees to which beings can commune, so construed, and (only) large differences ground differential degrees of moral status. The most glaring difference between beings in terms of their capacity to relate communally concerns whether they can be subjects or merely objects of such a relation. Having the ability to be a subject of communal relationship means being able to *commune with others*, that is, being able to think of oneself as bound up with others, to cooperate with others, to make their lives get better, and to do so out of sympathy and for their sake. In contrast, the ability to be an object of communal relationship means being able to *be communed with by others*. It means that an entity is such that typical human beings, by their nature and its, could think of themselves as bound up with it, promote its aims, help it, and do so out of sympathy and for its sake.

Consider, now, some of what this account of moral status entails. A very large majority of human beings can be subjects of a communal relationship, that is, they can commune with others, and (nearly) all living human beings can be objects of it, too (though perhaps not the permanently comatose). As typical human beings have the capacity to be both a subject and an object of communal relationship, they have the highest moral status, at least of beings on earth. I welcome the claim that they in fact have a dignity.

There is, of course, some debate about whether animals such as chimpanzees, elephants and dolphins can be subjects of communal relationship. If they can, it is probably to much less of a degree than typical human beings, and so they would have a somewhat lower moral status than humans. However, they would be rightly deemed to be 'higher' than other members of the animal kingdom, many of whom can be merely objects of communal relationship with us. Those animals capable of being only objects of communal relationship, such as cats, mice and fish, do have a moral status, albeit one lower than typical human beings and whichever animals might be capable of being subjects of communion.

There are some animals, however, that altogether lack a moral status by the present theory, roughly, those without a teleological orientation and a good, probably insects and bacteria. Since these beings have no aims with which to coordinate and no welfare to protect, we cannot commune with them in these ways, let alone sympathise with them or enjoy a sense of togetherness with them. It does not immediately follow that one may treat them however one likes, but any moral reasons for not harming them will be indirect.[2]

This African Modal Relational account of moral status naturally grounds certain judgements about how one is required to treat animals. The urgent interests of a being with moral status should not be traded off for the trivial interests of another being, even one with a higher moral status. And, so, if the urgent interests of animals that can be objects of communal relationship with us are at stake, it would be wrong to sacrifice them to satisfy our trivial interests. Such an approach forbids a wide array of practices that those who believe in animal rights will deem to be immoral, such as hunting for sport, eating meat for the taste and testing cosmetic products on animals.

However, Horsthemke is not satisfied with my theory. He does not question its African status, but rather maintains that its African status renders it unable to account adequately for animal rights.

Considering Horsthemke's objections to African Modal Relationalism

Horsthemke's three major objections are that any African theory, including the one I advance, is necessarily anthropocentric or speciesist, that my view is poorly defended, and that I have misapplied it, such that it entails that some animals have a moral status comparable to that of typical human beings. I address each point in turn.

According to Horsthemke, African ethics in general is essentially anthropocentric, meaning that I have to choose between the Africanness of my view and its ability to avoid anthropocentrism. He makes two claims in support of this claim. First, he notes, correctly, that one of the most influential ways that traditional Africans sum up their morality is with the phrase 'A person is a person through other persons' (for discussion in the context of several sub-Saharan peoples, see Nkulu-N'Sengha, 2009). To English-speakers unfamiliar with African worldviews, this maxim says very little. However, it may be plausibly interpreted as indicating (amongst other things) that one ought to become a real person, i.e., develop human excellence or what is called '*ubuntu*' in southern Africa, which one can do by relating communally with other persons. Horsthemke contends that, by this maxim, it is only through other *persons* that one can develop real or complete personhood, i.e., moral virtue. 'In focusing exclusively on human beings, *ubuntu* is by definition anthropocentric' (Horsthemke, 2015, p. 11; see also pp. 82–83, 93).

However, this maxim can be *essential* to (or typical of) an African ethic without being *exhaustive* of it. Adherents to such an ethic would typically say that someone lacked virtue if he tortured animals for fun, and often enough because of the pain inflicted on the animal, viz., for direct reasons. Although friends of a characteristically African ethic would say that more personhood or virtue would come from treating human beings well, or that it would be more base to mistreat human beings, they are not, merely by virtue of the maxim Horsthemke cites, committed to thinking that *only* human beings matter for their own sake, which I take to be the defining feature of anthropocentrism as it figures into philosophical ethics.[3]

Horsthemke's other reason for thinking that African ethics is essentially anthropocentric is that, as he again correctly notes, many traditional sub-Saharan societies call those who fail to develop personhood, and instead act immorally, 'animals' (or also 'non-persons,' on which see Nkulu-N'Sengha, 2009). According to Horsthemke, this suggests that animals lack moral worth (2015, p. 85).

However, to say that someone is 'an animal,' or, more carefully, is acting like one, is a metaphor indicating that the agent has acted morally *wrongly* (Ramose, 1999, p. 53; Gyekye, 2010). It does not imply anything about moral *status*, i.e., about whether animals warrant moral treatment for their own sake or not.

In reply, Horsthemke may point out that associating wrongful behaviour with animality suggests that those in such a culture deem the latter to be lowly. That is true. But it does not follow that anthropocentrism, understood as the view that only human beings merit moral treatment, is essential to their culture.

Turning to African Modal Relationalism in particular, Horsthemke points out that an implication of it is that, in cases where the urgent interests of animals and human persons conflict, the latter morally should win out. Although the trivial interests of persons do not outweigh the urgent ones of animals, by my view, it is true that, when a choice must be made between the urgent interests of two beings, the being with a greater moral status should prevail. Horsthemke calls this view 'necessarily anthropocentric' and also 'speciesist' (2015, p. 91).

However, African Modal Relationalism is not anthropocentric in the usual sense of viewing all non-human beings as lacking a moral status and having a merely instrumental value for human benefit. The view instead accords many animals a moral status and entails that it would be wrong to mistreat them in myriad ways that they are wronged in modern, industrialised societies and elsewhere.

Furthermore, African Modal Relationalism is not speciesist, because the reason for favouring human beings, in cases where their urgent interests are stake, is not the mere fact that they are human. Unlike the claim that it 'is with regard to their essential humanity that human beings matter' (Horsthemke, 2015, p. 5), by my view it is rather the fact that typical human beings have certain abilities to commune as subjects, that they can, roughly speaking, love. This is a capacity-based view, not one grounded on the property of humanness, and the capacity could well be shared by other beings. Indeed, it appears that chimpanzees, elephants and dolphins might share it to some degree.

In his book, Horsthemke does not discuss the strong intuitions that have led me to construct a theory with the implication that, in cases of forced trade-offs amongst the urgent interests of two beings, the one with the higher moral status should be favoured. There are several cases I have discussed in my work where it appears wrong to forsake a (human) person's life for that of a (non-personal) animal. For just one example, 'if the reader and I were starving along with a pig, and if I had a gun, it would be permissible for me to shoot the pig to feed us, but impermissible for me to shoot the reader to feed myself and the pig. The best explanation of such a judgement, I submit, is that pigs have a lower moral status than persons' (Metz, 2010, p. 309; see also Metz, 2012, p. 389).

For Horsthemke to point out that humans and animals can be equal but '*not* the same in all respects' (2015, p. 91) does not ground a clear answer to the question of how to choose between human and animal lives, in those unfortunate cases where a choice must be made. And the fact that such trade-offs might be rare is philosophically irrelevant, since the point is that consideration of such possible worlds reveals something about our implicit values. After all, speciesism is often questioned deeply on the ground that a race of intelligent and sentient Martians would have a moral status, even though we have never encountered such beings and never will.

African Modal Relationalism clearly forbids sacrificing urgent animal interests, such as life and freedom from excruciating pain, for at least the trivial interests of human beings, and so grounds vegetarianism, much anti-vivisectionist activism and related animal rights practices. So, I submit that it is not fair to say that my account with an African pedigree 'ultimately constitutes little improvement on traditional views, in its perspective on nonhuman animals' (Horsthemke, 2015, p. 92).

Horsthemke's second major criticism of African Modal Relationalism is that it is poorly defended, at least relative to his more egalitarian and individualist approach to moral status. One consideration here is whether it is plausible to think that a certain entity has a moral status in virtue of relational or extrinsic factors: 'How can "some kind of interactive property between one entity and another" . . . bestow status on the entities in question and not only on the relation between them?' (Horsthemke, 2015, p. 89; see also p. 99).

In reply, there is, upon reflection, little (if anything) odd or objectionable about saying that what makes a being particularly valuable and worthy of moral treatment is its natural capacity to relate to others in certain ways. Consider that what I have labelled a 'communal' relationship is one that many English-speakers would describe as a broad sense of 'love' or 'friendliness.' To love or to be friendly with others is, in large part, a matter of enjoying a sense of togetherness with them, coordinating one's behaviour with theirs, doing what one can to make their lives go well, and doing so consequent to sympathy and for the other's sake. It is not strange to suggest that a being might be important in virtue of its capacity to love as well as its capacity to be loved. Such a view implies that an entity's *relational properties* are what make it morally valuable, and do not push one to suggest that it is merely the loving *relationship* that is.

Horsthemke suggests that such a relational theoretical orientation is akin to objectionably reducing a person to a 'mere set of extrinsic relations' (2015, p. 89). According to him, my view does not even get the worth of persons correct, since they have an intrinsic value, i.e., a value based on their non-relational properties, and are not to be considered valuable merely insofar as they bear on others.

In reply, I note that, for one, there is a significant difference between a modal-relational theory and a corporatist one. According to the latter, only groups matter morally, with individuals serving as mere means, whereas the former grounds moral status on individuals' capacity to relate in principle, and not their actual relations. Although African Modal Relationalism rejects the idea that people's dignity is a function of properties that make no essential reference to others, it can coherently entail that people have a dignity, i.e., are good for their own sake to a superlative degree, by virtue of an ability to commune that by nature inheres in them.[4]

For another, I presume Horsthemke believes that animal *species* matter to some degree; if one were rare, it would deserve more protection, and not merely for ecological considerations of usefulness to other animals. If so, then *one* value of an entity can be *qua* member of a group. Though such does *not exhaust* an entity's value, this example shows that moral value is plausibly *not merely* a function of a being's intrinsic properties, opening up the door for a relational account.

Yet another reason Horsthemke finds African Modal Relationalism implausible is that 'if moral status is a matter of degree, then . . . a moral hierarchy could arguably be erected even within the realm of "normal adult humans": for example, among those who contribute more to communal well-being and life, who engage more readily in friendly, loving and trusting relationships than others' (2015, p. 89).

The straightforward way for me to reply is to point out that it is the *capacity* to commune, and not actual communion, that grounds moral status in my view. As it stands, this objection does not target African *Modal* Relationalism.

Perhaps, though, Horsthemke would reformulate the objection to contend that those by nature able to commune more would counterintuitively have a higher moral status. Now, on this point, I maintain that degree of moral status does not track degree of capacity to commune in a one-to-one fashion (see Metz, 2012, pp. 394–395). Only *large*, and not merely incremental, differences in degrees of natural ability to be either a subject or an object of a communal relationship constitute differences in moral status. Since there are in fact *not* such great differences among normal adult humans, this particular charge of hierarchy does not apply.

Were a stereotypical Jesus or angel to come on the scene, however, then the theory would be straightforwardly read as entailing that he would have a greater moral status than us. But, then, many would be prepared to bite the bullet in such a case. And for those not so prepared, it would not be *ad hoc* to follow most theorists of dignity in maintaining that it is what is sometimes called a 'range property,' such that any being with *enough* of a certain property has an equal standing relative to another.

At one point, Horsthemke questions my methodology, the way I argue for my theory of moral status. He notes, 'That it is a "persistent intuition" that "human beings have a greater moral status than animals" . . . does not establish much. Many of us have persistent intuitions about all kinds of things, including God's existence, ancestor agency and so on – which does not go anywhere near establishing the correctness of any of this' (2015, p. 91).

This point of course conflicts with Horsthemke's own recurrent appeals to intuition – indeed, as evidence against my theory – which are, I submit, unavoidable when doing moral philosophy systematically. More deeply, my *central* intuition was not one of unequal standing between animals and humans; instead, the appeals to intuition doing the most justificatory work were of specific cases in which one faces a forced choice between the lives of animals and humans, such as the pig case above. Those are more widely shared and difficult to give up, and are *best explained*, I submit, by a thesis of unequal standing. They therefore provide strong evidence (though of course no guarantee of truth) of differential moral status.

Horsthemke's third and final major objection is that, even if African Modal Relationalism's account of moral status were plausible, I have applied it incorrectly, failing to recognise that it actually entails that several animals have a moral status equal to ours. Even if one took my theory of moral status for granted, Horsthemke maintains that it would not entail that humans invariably have a greater moral status than animals. "(A)pes and monkeys, dogs, dolphins and whales, elephants, horses and birds like crows and jays 'are able to coordinate their actions with others, to do what is likely to make others better off, to act for the sake of others, and so on,' to use Metz's own description of higher autistic humans" (Horsthemke, 2015, p. 90).

Of course, the more Horsthemke is correct that my theory entails comparable moral status between human beings and animals, the less powerful are his previous objections to it on grounds of speciesism and the like. There is some tension between the sorts of criticisms Horsthemke makes of African Modal Relationalism.

Setting that point aside, though, my main reply is to note that Horsthemke's empirical claims are false or dubious, for all I know of the literature on animal cognition (and human autism) and *given the way I define these terms*. I am convinced that some animals have been shown to have a sense of self, an inner sense that is contrasted with an outer one. However, *having a sense of an other*, or, more carefully, *an other's mind*, is something else and has yet to be demonstrated, where this property is crucial for being a subject of communion as I construe it. It is very unlikely that dogs and birds can think about what others think about. And if chimpanzees can, then I would be glad to acknowledge their standing above everything else in the animal kingdom as, roughly, being able to love.

Conclusion: The nub of the debate

I developed African Modal Relationalism because I wanted a theory that was not merely African, and so grounded on ideas about hierarchy, relationality and community, but also attractive to a wide array of moral philosophers and ethically thoughtful people more generally around the world. The key judgements that drove me to develop a theory that entails differential moral status have been those in which it would be wrong to randomise when having to choose between the lives of many animals and persons (as well as those in which it would be wrong to randomise when having to choose between the lives of many animals and plants). Such intuitions, in which human lives should often be prioritised before those of animals (and in which the lives of animals should often be prioritised before those of plants), are plausibly explained by the thought that there is a hierarchy of moral standing. I have sought to cash out this hierarchy in terms of degrees of the visible (or 'physical') ability to be party to a communal relationship.

Horsthemke would give me the most reason to reconsider African Modal Relationalism if he were able to provide a satisfactory explanation of how morally to handle cases of forced trade-offs amongst the urgent interests of animals and human persons. Surely, he would not want to flip a coin when determining how to deal with the pig scenario above. Where there is a forced choice to make between feeding a human to a pig or a pig to a human, how are we to decide in a morally defensible way? Appealing to intrinsic properties such as being sentient, being able to flourish, being alive or a subject of life grounds an egalitarianism that is too strict for entailing that one ought to randomise which life to sacrifice. Of theoretical approaches to moral status, the idea that there is a hierarchy, and that the hierarchy is a function of the ability to commune, seems to me the most promising way philosophically to ground animal rights, while acknowledging their intuitive limits.[5]

Notes

1 This essay was first published in the *Journal of Animal Ethics* vol. 7, no. 2 (2017), and is reprinted here with some minor changes and with permission from the University of Illinois Press. Copyright 2017 by the Board of Trustees of the University of Illinois.

2 There are additional, and in some ways even more controversial, implications of my theory of moral status, not central to the debate with Horsthemke. For instance, certain animals are probably much more able to be communed with by characteristic human beings than others, e.g., cats versus fish, meaning the former have a greater moral status. In addition, 'marginal' cases of human beings are, I maintain, noticeably more able to be communed with by characteristic human beings than animals, meaning the former have a greater moral status. For discussion of these points see Metz (2012).

3 Horsthemke sometimes has a broader, and vaguer, notion of anthropocentrism, according to which it means 'human-centred' or involves assigning 'special value or worth to human beings' (2015, p. 5).

4 Compare this account of dignity with the discussion of 'non-intrinsic final goodness' common in the value-theoretic literature, e.g., Korsgaard (1983); Kagan (1998); Rabinowicz and Rønnow-Rasmussen (2000).

5 This work has benefited from the input of Kai Horsthemke and other participants at the Conference on African Philosophy: Past, Present and Future held at the University of the Witwatersrand, Johannesburg in 2015.

References

Behrens, K. (2010). Exploring African Holism With Respect to the Environment. *Environmental Values*, 19, 465–484.

Behrens, K. (2011). *African Philosophy, Thought and Practice and Their Contribution to Environmental Ethics*. Unpublished doctoral thesis, University of Johannesburg.

Behrens, K. (2014). An African Relational Environmentalism and Moral Considerability. *Environmental Ethics*, 36, 63–82.

Callicott, J. B. (1994). *Earth's Insights: A Survey of Ecological Ethics From the Mediterranean Basin to the Australian Outback*. Berkeley, CA: University of California Press.

Chemhuru, M. (2016). Using the African Teleological View of Existence to Interpret Environmental Ethics. *Philosophia Africana*, 18(1), 41–54.

Chitando, E., Adogame, A., and Bateye, B. (Eds.) (2012). *African Traditions in the Study of Religion in Africa*. Farnham: Ashgate.

Christians, C. (2004). *Ubuntu* and Communitarianism in Media Ethics. *Ecquid Novi*, 25, 235–256.

Gyekye, K. (2010). African Ethics. In E. Zalta (Ed.), *Stanford Encyclopedia of Ethics*. http://plato.stanford.edu/entries/african-ethics/.

Horsthemke, K. (2015). *Animals and African Ethics*. New York: Palgrave Macmillan.

Kagan, S. (1998). Rethinking Intrinsic Value. *The Journal of Ethics*, 2, 277–297.

Kasenene, P. (2000). African Ethical Theory and the Four Principles. In R. Veatch (Ed.), *Cross-Cultural Perspectives in Medical Ethics* (pp. 347–357). Sudbury, MA: Jones & Bartlett.

Kelbessa, W. (2011). *Indigenous and Modern Environmental Ethics: A Study of the Indigenous Oromo Environmental Ethic and Modern Issues of Environment and Development*. Washington, DC: The Council for Research in Values and Philosophy.

Korsgaard, C. (1983). Two Distinctions in Goodness. *The Philosophical Review*, 92, 169–195.

LenkaBula, P. (2008). Beyond Anthropocentricity: *Botho/Ubuntu* and the Quest for Economic and Ecological Justice in Africa. *Religion and Theology*, 15, 375–394.

Metz, T. (2010). Animal Rights and the Interpretation of the South African Constitution. *Southern African Public Law*, 25, 301–311.

Metz, T. (2012). An African Theory of Moral Status: A Relational Alternative to Individualism and Holism. *Ethical Theory and Moral Practice*, 15, 387–402.

Metz, T. (2013). The Western Ethic of Care or an Afro-Communitarian Ethic? Finding the Right Relational Morality. *Journal of Global Ethics*, 9, 77–92.

Metz, T., and Miller, S. C. (2016). Relational Ethics. In H. LaFollette (Ed.), *The International Encyclopedia of Ethics* (pp. 1–10). Malden, MA: Blackwell Publishing Ltd.

Molefe, M. (2015). A Rejection of Humanism in African Moral Tradition. *Theoria*, 62, 59–77.

Murove, M. F. (2007). The Shona Ethic of *Ukama* With Reference to the Immortality of Values. *Mankind Quarterly*, 48, 179–189.

Nkulu-N'Sengha, M. (2009). Bumuntu. In M. K. Asante and A. Mazama (Eds.), *Encyclopedia of African Religion* (pp. 142–147). Los Angeles, CA: Sage.

Odimegwu, I. (Ed.) (2007). *Perspectives on African Communalism*. Victoria: Trafford Publishing.

Olupona, J. (Ed.) (1991). *African Traditional Religions in Contemporary Society*. New York: Paragon House.

Rabinowicz, W., and Rønnow-Rasmussen, T. (2000). A Distinction in Value: Intrinsic and for Its Own Sake. *Proceedings of the Aristotelian Society*, 100, 33–51.

Ramose, M. (1999). *African Philosophy Through Ubuntu*. Harare: Mond Books.

Teffo, L. J., and Roux, A. (1998). Themes in African Metaphysics. In P. Coetzee and A. Roux (Eds.), *The African Philosophy Reader* (2nd ed., pp. 192–208). London: Routledge.

Wiredu, K. (2008). Social Philosophy in Postcolonial Africa: Some Preliminaries Concerning Communalism and Communitarianism. *South African Journal of Philosophy*, 27, 332–339.

4 An African account of the moral obligation to preserve biodiversity

Kevin G. Behrens

Introduction

Scientists define a mass extinction as an event in which more than 75% of the planet's species are lost over a relatively short period of time, geologically speaking. Five such mass extinctions have occurred in history, each coinciding with the end of a major geological period: the Ordovician, Devonian, Permian, Triassic and Cretaceous periods (Barnosky et al., 2011, p. 51). It is now commonly claimed that the earth is entering a sixth mass extinction. In a recent study Ceballos et al. claim that we have lost vertebrate species at an average of up to 100 times the background rate in the last century – and this is based on a very conservative estimate of the natural background extinction rate. This is evidence enough for them to conclude that a new mass extinction event has indeed already started happening (Ceballos et al., 2015). Of course, the loss of species is common and part of a natural rhythm. It is estimated that some four billion species have evolved in the earth's history, and that 99% of those no longer exist. What is different about our current age is that extinctions are not in balance with new speciation (Barnosky et al., 2011, p. 51). The huge acceleration in species loss results in an increasing net reduction in biodiversity. Barnosky et al. sum up the main concerns about biodiversity loss:

> humans are now causing the sixth mass extinction, through co-opting resources, fragmenting habitats, introducing non-native species, spreading pathogens, killing species directly, and changing global climate. If so, recovery of biodiversity will not occur on any timeframe meaningful to people: evolution of new species typically takes at least hundreds of thousands of years, and recovery from mass extinction episodes probably occurs on timescales encompassing millions of years.
>
> (Barnosky et al., 2011, p. 51)

While there is significant agreement that the huge acceleration in loss of biodiversity is primarily caused by human activity, it is not obvious that we are morally obliged to prevent species loss. Environmental ethicists have struggled to provide a robust account of our moral obligations towards species *qua* species, rather than merely towards individual members of species. I argue that there are notions in African thought that can enrich our discourse about obligations to preserve biodiversity. Firstly, I argue that there are African conceptions that are able to overcome the challenge of anthropocentrism – the idea that nature has no intrinsic value, and has only instrumental value to humans. I show that many African theorists point to the interrelatedness of natural entities as grounds for treating nature with respect. I also argue that a sub-Saharan conception of moral considerability, grounded in the notion of an interconnected "web of life," is able to accord moral standing to species *qua* species. Secondly, I give an account of an African conceptualisation of our moral duties to future generations that entails a strong obligation to preserve biodiversity for posterity. I identify two salient notions in African thought that ground an obligation to protect nature for the sake of future generations. The first is that the natural environment does not belong to the present generation, but is shared across generations. The second is that the present generation owes a debt of gratitude to ancestral generations for having preserved the environment on its behalf, and that this gratitude ought to motivate the present generation to protect nature for the sake of posterity, too. Finally, I argue – pragmatically – that, given the value of eco-tourism for African countries, the preservation of the rich biodiversity of the continent ought to be a priority.

I want to quickly set aside the claim that species loss is inevitable and simply part of the cycle of evolution, and therefore not something we should have any moral responsibility for. It is true, of course, that there would be extinctions on earth, even without humans. In fact, the five mass extinctions that have occurred in the earth's past all occurred before the evolution of our species. Certainly, to the extent that species would be lost without the presence of humans, and in cases where the loss of certain species now might be unrelated to human activity, we clearly are not morally accountable. However, what does impose a moral obligation on us with respect to the biodiversity loss of our time is that it is primarily our behaviour that is causing it, and that the magnitude of the loss is so much greater than what would be occurring without our influence. Nonetheless, even if it is granted that we are the main cause of the acceleration in species loss, accounting for any purported moral obligation to prevent it is notoriously difficult.

The challenge of anthropocentrism

A major challenge to the notion that we can have a moral obligation to preserve biodiversity is the challenge of anthropocentrism. Since its earliest beginnings as a new field of applied ethics, environmental ethics has been characterised by a strong rejection of the anthropocentric assumptions of most moral philosophy, historically. A central claim made by many environmental philosophers is that the roots of the current ecological crisis lie in the dominance of anthropocentric views, particularly in the developed Western world. Many environmental ethicists have laid the blame for environmental damage squarely at the door of a moral value system that attributes final moral value only to humans (J. Callicott, 1986; Rolston, 1975; Routley, 1973; White, 1967). All other entities, including other living beings, are morally valued only in terms of their usefulness to humans. This strong anthropocentrism gives rise to the notion that humanity is free to exploit and even abuse nature in any way it likes, as nature's only value lies in its instrumental value to us. The challenge of anthropocentrism to the idea that we have a moral obligation to preserve biodiversity is clear. If the claims of the anthropocentrists are right, and the only entities that are of intrinsic moral value are humans, then we have an obligation to preserve biodiversity only to the extent that the preservation of any particular species is instrumentally important to humans. Thus, it really would not matter if we were to drive a large percentage of species to extinction, as long as their disappearance would not have a negative impact on human well-being. Clearly, such an approach is not satisfactory to those who want to defend the notion that we have an obligation to preserve biodiversity.

Environmental ethicists have, for more than half a century, been trying to develop moral theories that reject strong anthropocentrism in favour of an acknowledgement of moral value in not just humans, but in other living beings, too. A vast body of literature now exists, written mainly by Western theorists, that supports the need to reject anthropocentrism. I think this body of work can be significantly strengthened by under-appreciated insights from African thinkers. For a variety of reasons ranging from simple ignorance to overt racism, the contribution that African thought can make to our environmental ethical discourse in general has largely been overlooked. Where African moral traditions and philosophical conceptions have been considered at all as sources for developing an environmental ethic, they have frequently been all but written off on the basis of an assumption, quite prevalent in the literature, that African thought is inherently anthropocentric. J. Baird Callicott notoriously claims that "Africa looms as a big blank spot on the world map of indigenous environmental ethics for a very good reason. African thought orbits, seemingly, around human interests" (Callicott, 1994, p. 158). Elsewhere, I have argued at some length that it is false to assume that African thought is inherently anthropocentric (Behrens, 2010). Of course, it is true that there is much anthropocentrism to be found in the work of African authors, but this is just as true of Western work. It is also clear that human interests feature prominently in African notions of moral obligation. However, the African literature on humans and their relationship with the environment is full of claims that humanity and all of the rest of nature are interrelated and interdependent. There is a clear recognition of the insights of modern ecology that the natural environment is an interconnected system. For many African theorists, this interrelatedness means that the well-being of humans is dependent on the well-being of other natural entities, and this imposes an obligation on us to treat nature with respect.

I briefly quote some representative examples of claims of this kind. Munyaradzi Felix Murove writes that "human well-being is indispensable from our dependence on and interdependence with all that exists, and particularly with the immediate environment on which all humanity depend" (Murove, 2004, pp. 195–196). Benezet Bujo writes: "The African is convinced that all things in the cosmos are interconnected. All natural forces depend on each other, so that human beings can live in harmony only *in* and *with* the whole of nature" (Bujo, 1998, pp. 22–23).

Godfrey Tangwa asserts: "The pre-colonial traditional African metaphysical outlook . . . impl[ies] recognition and acceptance of interdependence and peaceful coexistence between earth, plants, animals and humans" (Tangwa, 2004, p. 389). Workineh Kelbessa, writing of the Oromo people's moral outlook claims that it ". . . does not allow irresponsible and unlimited exploitation of resources and human beings . . . It reflects deep respect and balance between various things. The Oromo do not simply consider justice, integrity and respect as human virtues applicable to human beings but they extend them to nonhuman species and mother Earth" (Kelbessa, 2005, p. 24). Kofi Opoku writes: "There is community with nature since man is part of nature and is expected to cooperate with it; and this sense of community with nature is often expressed in terms of identity and kinship, friendliness and respect" (Opoku, 1993, p. 77). It is obvious that African thought is not hopelessly anthropocentric, and that on many accounts, there is a requirement to treat nature with respect because of a fundamental interdependence of all natural entities.

What I have suggested thus far is that there are notions about our relationship with nature to be found in the work of African theorists that entail that we have an obligation to value nature, not just for its instrumental value to humans, but in and of itself. What is not yet clear is exactly what sorts of entities other than humans ought to be accorded some value and on what fundamental basis. The essential question I am addressing here, is exactly what entities can be said to be morally considerable. By morally considerable I mean the capability to be the kind of thing towards which we humans can have any kind of direct moral obligation at all. Thus, if an entity is morally considerable, it must have some moral value other than merely its instrumental value for us as humans. Among the group of morally considerable entities, there will clearly be vast differences in actual moral status. These differences in status will result in some kind of hierarchy in terms of which entities should be prioritised when our duties towards them come into conflict. But this is beyond the scope of this article. All I need to show here is that the entities that may be said to belong to natural species qualify as morally considerable.

There has been much debate in the existing (largely Western) literature about what criteria may be used to establish the moral considerability of entities. There are those who propose (variously) that the property that grants moral considerability is consciousness, self awareness (Tooley, 1976) sentience (Singer, 1975), rationality (Fletcher, 1966) and a whole list of others. I argue for a criterion for establishing moral considerability that is grounded in the African notion of the interrelatedness of all natural entities, and more specifically in the notion of an interconnected web of life.[1] A recurring image in the African literature is of life itself being described as a web, network, texture or fabric. What is rejected is the familiar Western notion of life as being something possessed by individual beings. Rather, it is suggested that life is shared, not owned. It is a web into which each of us connects. Oruka and Juma speak of "pan-organism as the basic truth underlying all nature . . . This means that the ecological activities are a network. So a break or imbalance in one aspect has serious consequences in other aspects of the domain" (Oruka and Juma, 1994, p. 122). Harvey Sindima writes that "nature and persons are one, woven by creation into one texture or fabric of life, a fabric or web characterized by an interdependence between all creatures" (Sindima, 1990, p. 143). He describes life not as something possessed by individual organisms but rather as a "woven . . . texture or fabric of life, a fabric or web characterized by interdependence" (Sindima, 1990, p. 143). On this understanding of life itself as a web or fabric, I argue that the property which grants moral considerability to any entity is that it is part of this web of life, and that all entities that are part of this web of life are morally considerable. Simply put, because all things that are part of the web of life are interconnected and their well-being in some way related, they are entities that need to be taken into consideration when making moral decisions.

Clearly, on this criterion, all living organisms are morally considerable because they are part of the web of life.[2] They may have very different degrees of moral status, but they are all entities that need to be taken directly into consideration in our moral deliberations. Since species are groups of closely genetically related living organisms, on a web of life centred notion of moral considerability, at least individually, members of natural species count morally, and not only in terms of their usefulness to humans. This allows for us to conceive of having an obligation to preserve biodiversity and to protect species even when their loss would have minimal consequences for human well-being.

The challenge of accounting for duties to species qua species

Even if a convincing case can be made that the individual members of species are morally considerable and that strong anthropocentrism is false, it still does not mean that we have an obligation to protect species as collectives or wholes, rather than just to take their individual members into account morally. In fact, accounting for moral obligations to species qua species has proven to be notoriously difficult. This is problematic, because unless we can give a coherent account of obligations to species qua species, we have very few grounds for justifying why we ought to take any special measures to protect species that are particularly endangered. Critics of individualist biocentrism have long highlighted this concern. As Lilly-Marlene Russouw puts it: "duties toward individual beings . . . will not always account for all the actions that people feel obligated to do for endangered species – e.g., bring into the world as many individuals of that species as possible, protect them from natural predation, or establish separate breeding colonies" (Russouw, 2003, p. 470). To support our general intuition that these are exactly the kind of obligations we have to endangered species, it is necessary to be able to theoretically defend the notion of having obligations to species qua species.

One of the reasons that it is difficult to talk of obligations toward species as species is that it is not easy to conceive of a species as having interests, other than in terms of the aggregated interests of its individual members. Much has been made of this problem in the Western literature (VanDeVeer and Pierce, 2003, pp. 452–453). Once again, I think African thought has much to contribute to resolving this problem. That there is a strong communitarian element to African notions of the moral life is a truism. Certainly, the Western preoccupation with atomistic individuals does not sit well with Africans generally. For many African groups, there is a sense in which individual persons obtain full personhood only in and through their relationships in community. Michael Eze writes: "the relationship between the individual and the community is dialogical for the identity of the individual and the community and is dependent on this constitutive formation. The individual is not prior to the community and neither is the community prior to the individual" (Eze, 2008, p. 386). Africans do place a great deal of emphasis on social groups, including the family, clan and tribe. There is a sense of individuals being deeply embedded in their social relationships. Since the individual is to some extent incomplete outside of these relationships, on this worldview, the Western scepticism about whether groups can have rights is nonsensical. On African communitarian thinking, the individuals who make up a family or clan are harmed when the group as a whole is harmed. There is no sense of questioning whether a family, clan or tribe is the kind of thing that can have interests or be harmed. Since a species is a similar sort of grouping to a human clan or tribe, African communitarians would find no theoretical problem with the idea that species qua species can have interests, and that we can correspondingly have obligations towards them.

Another serious threat to the notion of obligations toward species as species is the fuzziness of the notion of a "species" itself. Is what we commonly refer to as a species a real natural entity, or is this classification of organisms merely arbitrary (VanDeVeer and Pierce, 2003, p. 456)? There has been much debate about this in the Western literature (Gould, 2003; Rolston, 2003; Russouw, 2003). According to Stephen Gould, the standard basic definition of species is "a population of actually or potentially reproducing organisms sharing a common gene pool" (Gould, 2003, p. 467). On this definition, the criterion for distinguishing species is the inability to breed with other species. This criterion also helps to deal with the problem of species in evolutionary transition and when we could say that they have crossed the line and become a new species. Gould claims the "reproductive criterion generally works well" (Gould, 2003, p. 467). However, he acknowledges that it does not always work, as there are exceptions in which interbreeding is possible across species lines. In fact, it turns out that there is no criterion for species definition that can be applied to all species without exception. Gould does not regard this as too much of a problem, as

> nature is nothing if not a domain of exceptions, where an example against any clean generality can always be found . . . But the criteria work in the vast majority of cases, including humans. Species are not arbitrary units, constructed for human convenience, in dividing continua. Species are the real and objective items of nature's morphology. They are 'out there' in the world as historically distinct and functionally separate populations 'with their own historical role and tendencies'.
>
> (Gould, 2003, p. 468)

Holmes Rolston III makes similar claims, and his contribution is worth quoting at some length:

> Species exist only as instantiated in individuals, yet they are as real as individual plants or animals. The assertion that there are specific forms of life historically maintained in their environments over time seems as certain as anything else we believe about the empirical world. At times biologists revise the theories and taxa with which they map these forms, but species are not so much like lines of latitude and longitude as like mountains and rivers, phenomena objectively there to be mapped. The edges of these natural kinds will sometimes be fuzzy, to some extent discretionary. One species will slide into another over evolutionary time. But it does not follow from the fact that speciation is sometimes in progress that species are merely made up and not found as evolutionary lines with identity in time as well as space.
>
> (Rolston, 2003, p. 477)

Perhaps Gould and Rolston give us sufficient grounds for thinking that the classification of species is not merely arbitrary or so fuzzy as to be meaningless. But what is the status of species as a category in African thought? There is much evidence that the concept of a species has been recognised by Africans for centuries, and that some notion of obligations towards certain species is fairly common and widespread in African communities. That species are recognised as a distinct category is most obvious in the pervasive existence of totemism in Africa. Many groups regard certain species as totems, often granting them a special status and a degree of protection. These totemic species are generally associated with a particular tribe or clan symbolically and harming members of such species is prohibited. In discussing the Shona conception of totemic species, Murove writes, "kinship constitutes the reason behind the Shona notion of totemic ancestorhood whereby the individual's historical origins and identity are traced to other species such as eland, kudu, fish, zebra, lion, crocodile, just to mention a few" (Murove, 2004, p. 185). Furthermore, some authors suggest that totemic (or some spiritually revered status) may be accorded not only to animal species, but to plant species as well (Kelbessa, 2005, p. 24).

Of course, establishing the fact that many different species are granted an elevated status and protection by groups across the African continent hardly, on its own, suggests that it is common for African communities to affirm an obligation to protect species in general. Indeed, as Kai Horsthemke points out "it turns out . . . that one clan's totem animal is another clan's favourite 'bush meat'" (p. 74). Nonetheless, what these beliefs do establish is that the notion of a species is recognised, and that it is not entirely foreign to conceive of obligations to protect species qua species. Interestingly, at least one author actually suggests that the implications of totemic and other beliefs entail that the destruction of species is a moral wrong. Writing of the Oromo people, Kelbessa writes:

> The Oromo do not merely focus on creatures that have economic importance but they also pay attention to other species as valuable in and of themselves.

> Sacred groves have symbolic meanings. Similarly, certain wild animals are looked upon as symbols of unity and have religious significance. . . . The important principle arising from Oromo wildlife management is that it is morally wrong to totally destroy a species and that humans should live in harmony with other creatures.
>
> (Kelbessa, 2005, p. 24)

I think the most compelling grounds for arguing that African thought affirms the notion of duties to species as species is hinted at by Placide Tempels, who writes, "an analogy can be found between a human and a lower group (e.g., in the animal class), an analogy based on the relative place of these groups in relation to its own class." The key to understanding why African thought recognises obligations to species might well lie in the simple but obvious observation that species are analogous to human groups, such as clans and tribes. As humans we identify with our social groups and often seek to look out for one another's interests. It is easy enough to recognise similar behaviour in other species, at least animal species. Furthermore, just as the need to procreate and perpetuate our own kind is hardwired into humans, it is surely quite obvious that other species, including plants, also seek to perpetuate their kind and ensure its continued existence. It is in a sense a kind of solidarity, which recognises the interest of other species to thrive and survive, that undergirds a sense of obligation towards other species qua species in African thought.

Moral obligations to future generations

I now turn to another challenge that environmental ethics has had to grapple with and that is of much relevance to the notion of a moral obligation to preserve biodiversity: the question of whether it is plausible to conceive of moral obligations to future generations.[3] The reason this is so important is that, with many environmental problems, the worst effects of human actions in the present will be felt only in generations to come, in some cases, many decades into the future. This is particularly true with respect to the environmental harms that will result from escalating global warming caused primarily by our continued releasing of greenhouse gasses into the atmosphere. It is now generally accepted that we are already beginning to experience the effects in terms of climate change, but that the most significant threat lies in far more dramatic changes in global climate that will occur only generations into the future, unless we change our behaviour now. Scientific modelling allows us to predict that climate change will have enormous implications for future generations of humans. It is also the single biggest threat to biodiversity in the future. Species loss will accelerate quite dramatically in a future increasingly affected by climate change.

However, it could be argued that it is not our moral responsibility to consider the effects of our conduct on future generations. And, indeed, unless we are able to give a robust account of exactly how it is plausible to conceive of obligations to posterity, we might be forced to concede that we have no need to concern ourselves with any potential loss of species or other environmental harms that our actions might give rise to in the more distant future. While I think it is true that many people have an intuition that we clearly do have an obligation to preserve the environment for posterity, in the Western literature, theorists have struggled to give a plausible account of this obligation. I have dealt with this issue in some depth in other work, where I have claimed that there are some very valuable insights in African thought that can make a significant contribution to providing solid theoretical grounding for duties to future generations (Behrens 2012). I briefly present my arguments in what follows.

Let me start with the theoretical problems regarding moral obligations to future generations that have been identified in the dominant Western discourse. Ernest Partridge, who published the first book-length account of duties to posterity points out that to the average person it seems intuitively obvious that we do have moral obligations to future generations. Yet, on deeper reflection this idea is beset with theoretical problems. He describes how philosophers have asked how it is possible for us to have moral obligations to:

- beings that don't yet exist
- whose very future existence is contingent on our choices now
- who can't reciprocate our consideration towards them
- who are not individually identifiable
- and whose needs/interests in the future cannot be anticipated now (Partridge, 2003).

These are not trivial concerns, and unless they can be answered plausibly and robustly, it seems that the claim that we can have obligations to future generations is in trouble. There has, therefore, been an extended academic debate around this claim in the Western literature.

By contrast, many African philosophers are confident in asserting moral obligations to posterity. Kwasi Wiredu writes:

> Of all the duties owed to the ancestors none is more imperious than that of husbanding the resources of the land so as to leave it in good shape for posterity. In this moral scheme the rights of the unborn play such a cardinal role that any traditional African would be non-plussed by the debate in Western philosophy as to the existence of such rights. In upshot there is a two-sided concept of stewardship in the management of the environment involving obligations to both ancestors and descendants which motivates environmental carefulness, all things being equal.
>
> (Wiredu, 1994, p. 46)

Wiredu clearly believes that we do have moral obligations to future generations, at least with respect to "leaving the land in good shape for posterity." He links this moral duty with a corresponding obligation to the ancestors. I unpack this idea in more detail later.

Wiredu is not alone among African theorists in linking an obligation to preserve the environment for future generations to beliefs about ancestors. Benezet Bujo describes the African "ethical community" (or, if you like, circle of moral obligation) as including the living, the ancestors and future generations (Bujo, 1998, p. 27). Felix Murove describes Bujo's sense of a moral community across generations as "anamnestic solidarity" – solidarity with future generations grounded in memories of the past:

> The life experience of the current generation is understood as being grounded in the past, and . . . the present will contribute to its own existence into the future when it has become the past. Through anamnestic solidarity, the communal life of the living (the present) and that of the ancestors (the past) is re-enacted as a gift to be shared and passed on to posterity.
>
> (Murove, 2004, p. 184)

He goes on:

> One should always live and behave in a way that maximizes harmonious existence in the present as well as in the future. In other words, our human behaviour at present that takes into account the concerns of the future generations becomes an investment in the future.
>
> (Murove, 2004, p. 184)

What is clear is that at least some prominent African philosophers do not share the doubt of many of those in the Western tradition that moral obligations to future generations can be plausibly upheld. Of course, many non-Africans will be incredulous about beliefs that the ancestors – the recently deceased – can continue to have a direct influence on the experiences of the living. However, I shall show that, while linked to contentious beliefs about the so-called living dead, these African notions that entail obligations to posterity can be set free from their traditional metaphysical moorings, secularised, if you like, and still provide cogent grounds for moral obligations to future generations.

The first salient African notion entailing moral duties to posterity that I discuss is related to the assertion that the land – broadly understood as the natural environment – is not something that can be fully owned or be the possession of the current generation. It is rather to be conceived of as a resource or inheritance that is rightfully shared across generations.

In much of the work of African theorists, the idea of the environment being shared across generations is understood in terms of the land belonging not to the current generation, but rather to the ancestors, who act as custodians of the land. But this idea does not rely on a belief that ancestors continue to influence the lives of the living for it to be plausible. It is a simple fact that each generation inherits an environment in a particular state from its predecessors, and it will pass it on in a particular state to those who will follow. The environment is a resource shared by successive generations, and as with any shared resource, those who make use of it have an obligation to others who use it to make fair use of it. It is analogous to any shared facility, for example a public restroom. This is a facility that is shared by a succession of users over time. And because the facility is shared, each user has an obligation to future users not to leave the facility in a dirty state or, more seriously, to vandalise it. The fact that each user does not know the exact identity of future users does not make this duty any less real. Nor does the fact that later users cannot reciprocate the consideration of previous users directly.

Thus, although some African philosophers associate the obligation to preserve the environment for future generations with a belief in ancestors, what really does the work in entailing this moral obligation is the fact that the environment is a resource shared across generations. Matthew Izibili likens each generation's possession of the environment to a mortgage, suggesting that the land is not so much owned as it is on loan to its current users (Izibili, 2005, p. 383). I have a strong suspicion that these ideas are not unique to Africa, and that they may be shared by many other cultural and religious communities around the world. Perhaps, then, it is mainly a Western anomaly to question obligations to posterity, and we should join Wiredu in being "non-plussed by the debate" in the Western literature (Wiredu, 1994, p. 46).

In support of this view, John O'Neill has argued that modern Western culture with its strong individualist bent and fierce attachment to the notion of private ownership has lost sight of the fact that nature is a resource shared by successive generations and that the circle of moral obligation crosses generations. He uses the metaphor of partial blindness to describe this phenomenon:

> a . . . temporal myopia . . . infects modern society. The question of obligations to future generations is posed in terms of abstract obligations to possible future people who are strangers to us. The argument is premised on the lack of a sense of continuity of the present with both the past and the future.
>
> (O'Neill, 1993, p. 47)

O'Neill claims that theoretical concerns about how moral obligations to future generations can be plausible arise because they are framed in individualist rather than communal terms. We can't know who future individuals will be and it is hard to conceive of claims that future, impersonal, unknown, contingent individuals could have on us. But, if we think in communal terms, whole generations, as collectives, can be seen to have a duty to preserve the environment for future generations as collectives. We don't need to know who the individuals will be to know that a future generation is inevitable (short of cataclysm). If the land is seen as a shared resource, each generation can be seen as having a duty to preserve it for the next – even if future individuals are contingent and unidentifiable.

The second African notion that is able to contribute to our understanding of duties to posterity relates to gratitude. Recall that Wiredu writes: "In upshot there is a two-sided concept of stewardship in the management of the environment involving obligations to both ancestors and descendants which motivates environmental carefulness, all things being equal" (Wiredu, 1994, p. 46). Unpacking this further, the basic idea seems to be that the current generation ought to have an attitude of gratitude to their predecessors for having preserved the environment for their sake. In turn, this gratitude ought to inspire the living to emulate the example of their predecessors and use the environment cautiously and with care for the sake of future generations. So, there is a sense of a debt of gratitude to our predecessors that needs to be paid, not to them, but to our descendants. Now, clearly this would work, inasmuch as if we did develop this attitude, we would do a much better job of protecting the environment than we are doing now. But does it make any sense? The problem, of course, is that it is difficult to conceive of how a duty of gratitude to one party can impose a moral obligation on us towards a different party. It seems somewhat counterintuitive.

Yet I think the idea may be more plausible than it seems at first. Firstly, it makes more sense when we conceive of gratitude less as an affective response and more as an attitude or general disposition – a virtue if you like. If our response to previous generations' care of the land for us is to develop a predisposition to gratitude, we may feel compelled to show similar kindness to another party. Michael McCullough writes:

> Like hope, one of the key psychological processes governing gratitude may be a mindful awareness – specifically, awareness of how one's life is held together through the benevolent actions of other people. Grateful people, on recalling a positive outcome in their lives, are mindful of the causal agents . . . who have acted in ways that benefitted them. . . . [G]rateful people . . . pay attention to the ways in which their lives are connected to other events and activities occurring in the social, natural, and (for some people) supernatural world.
>
> (McCullough, 2002, p. 303)

So, perhaps an attitude of gratefulness can inspire in us a disposition to be benevolent towards others, even future generations.

Yet, even if the virtue of gratefulness could inspire a desire or disposition in us to show favour to others – it would still not constitute a *moral obligation* to care for others. Is it reasonable to think that gratitude to one party can impose a duty on us to care for another?

Claudia Card writes with respect to responsibilities of gratitude, "Some responsibilities are naturally carried out in relation to still other persons: a hitchhiker may pick up future hitchhikers, for example, or a student give help and encouragement to newer students" (Card, 1988, p. 120). Consider mentorship in academia: it not unusual for us to respect the efforts of those who have mentored us by seeking to invest our own energy in the next generation of scholars. We can't ever repay our mentors, but we can honour them by emulating them and helping our students. So, perhaps it is not so unreasonable to think that an attitude or virtue of gratefulness to our predecessors for preserving the land for us might impose on us a duty or responsibility to future generations to ensure that we leave them an environment in good shape, including rich biodiversity.

Thus, I conclude that, on the basis of some salient African notions, grounded in, but not dependent upon, beliefs about ancestors, a very plausible account can be given of why we do have moral obligations towards future generations. These obligations primarily relate to the natural environment, including the diversity of species on earth. Specifically, our obligation to posterity is to leave them a biosphere that is as far as possible as rich, fertile, plentiful and sustaining as the one we inherited from our ancestors.

The value of eco-tourism as a reason for preserving biodiversity

I wish to briefly make one final argument for a moral obligation to preserve biodiversity. This argument is essentially pragmatic in nature, inasmuch as I want to defend the view that one of Africa's greatest resources is its natural environment, including its wilderness areas and its exceptional biodiversity. There is a growing global market in eco-tourism that can be tapped by African countries far more than is the case at present. Moreover, as long as this natural resource is maintained at sustainable levels, it is in a sense renewable, as generation after generation of natural organisms replace their predecessors. If we are careful, then long after all of the oil has been tapped, the minerals exhausted and metals extracted, our continent will still have the riches of its biotic heritage. These are resources that with careful stewardship can still be of benefit to our descendants for many generations to come. This seems to me as good a reason as any for the preservation of biodiversity to be a major priority for the sake of the people of this continent.

Of course Africa is not unique in being able to boast of its enormous biodiversity. There are other regions of the world that also have much to offer eco-tourists. What is unique about Africa is something about the precise nature of its diversity of species. In particular, Africa is blessed with a significant variety of what are known as "charismatic megafauna." These are typically animal species – usually, but not always – fairly large, that are particularly appealing to the public. These are species that conservationists and eco-tourism marketing agencies hold up as the "poster kids" for promoting their causes or interests (Hausmann et al., 2016). It is not for nothing that World Wildlife Fund's trademark emblem is a giant panda. It is hard to imagine a cuter mascot to use to draw attention to the plight of endangered species. Callicott beautifully reminds us of Africa's rich heritage when it comes to these kinds of species:

> tropical Africa is the richest place on earth for what conservationists call 'charismatic megafauna.' Indeed the mere mention of Africa conjures images in the mind's eye of wildebeests, springboks, hippopotami, rhinoceroses, zebras, giraffes, elephants, ostriches, flamingos, crocodiles, lions, leopards, cheetahs, monkeys, baboons, gorillas, chimpanzees, and many other kinds of animals.
>
> (Callicott, 1994, p. 156)

Conservationists have long understood that these species are marketing "eye candy," the tempting bait with which to lure potential donors to part with their cash. Promoters of eco-tourism are also aware that it is the promise that tourists will be able to see many of these sorts of species in one place that will draw larger numbers of visitors. Just consider the number of parks and reserves that boast that they host the "big five," as an example of how these charismatic megafauna are able to attract tourists. It seems to me that this is a very special asset, and one that our continent can ill afford not to make the most of.

I am not at all unconscious of the concern that the focus on popular and charismatic species might draw attention away from less attractive but equally important and possibly more endangered species, including plants. However, there is evidence that using these charismatic species as drawcards has been quite successful in acting as a gateway to draw attention (and funding) to the plight of other less popular species (Hausmann et al., 2016). Besides, I am not in any way trying to justify the advocacy or marketing methods of conservation bodies or the tourism industry. I am merely highlighting the fact that the biodiversity of Africa is of such a nature as to make it a sustainable and extremely valuable renewable source of important revenue for the continent long into the future. This is an exceptionally good reason for us to preserve biodiversity.

Conclusion

In a context in which human activity is the main cause of an alarming increase in the number of species becoming extinct, many hold the view that we have a moral obligation to try to prevent extinctions and to preserve biodiversity. Providing a robust account for such obligations has proven to be quite difficult. I have argued that there are important notions in African moral thought about humans and their relationship with nature and about the possibility of moral duties to future generations that can contribute in meaningful ways to developing a plausible account of the moral obligation to preserve species and biodiversity.

Notes

1 I provide a much fuller description of an African "web of life" account of moral considerability in (Behrens, 2014). I also make a case for the plausibility and attractiveness of such an account in this article.

2 On this criterion, it is also possible to grant moral considerability to certain non-living natural entities such rivers, mountains and even rock formations, as well as certain artefacts. See (Behrens, 2014).

3 I give a very full account of moral obligation to future generations grounded in African thought in (Behrens, 2012). Much of the account given in this paper is based on this article.

References

Barnosky, A. D., Matzke, N., Tomiya, S., Wogan, G. O. U., Swartz, B., Quental, T. B., . . . Ferrer, E. A. (2011). Has the Earth's Sixth Mass Extinction Already Arrived? *Nature*, 471, 51–57. http://doi.org/10.1038/nature09678

Behrens, K. (2012). Moral Obligations Towards Future Generations in African Thought. *Journal of Global Ethics*, 8(2/3),179–191.

Behrens, K. (2014). African Relational Environmentalism and Moral Considerability. *Environmental Ethics*, 36(1), 63–82.
Bujo, B. (1998). *The Ethical Dimension of Community*. Nairobi: Paulines Publications.
Callicott, J. (1986). The Search for an Environmental Ethic. In T. Regan (Ed.), *Matters of Life and Death: New Introductory Essays in Moral Philosophy*. New York: Random.
Callicott, J. B. (1994). *Earth's Insights: A Survey of Ecological Ethics From the Mediterranean Basin to the Australian Outback*. Berkley, CA: University of California Press.
Card, C. (1988). Gratitude and Obligation. *American Philosophical Quarterly*, 25, 115–127.
Ceballos, G., Ehrlich, P. R., Barnosky, A. D., García, A., Pringle, R. M., and Palmer, T. M. (2015). Accelerated Modern Human – Induced Species Losses: Entering the Sixth Mass Extinction. *Sciences Advances*, 1(e1400253), 1–5. http://doi.org/10.1126/sciadv.1400253
Eze, M. (2008). What Is African Communitarianism? *South African Journal of Philosophy*, 27, 386–399.
Fletcher, J. (1966). *Situation Ethics: The New Morality*. Philadelphia, PA: Westminster Press.
Gould, S. (2003). What Is a Species? In D. VanDeVeer and C. Pierce (Eds.), *The Environmental Ethics and Policy Book* (3rd ed.). Belmont, CA: Wadsworth.
Hausmann, A., Slowtow, R., Fraser, I., and Di Minin, E. (2016). Ecotourism Marketing Alternative to Charismatic Megafauna Can Also Support Biodiversity Conservation. *Animal Conservation*. 20(1): 91–100. http://doi.org/10.1111/acv.12292
Izibili, M. (2005). Environmental Ethics: An Urgent Imperative. In P. Iroegbu and A. Echekwube (Eds.), *Kpim of Morality Ethics*. Ibadan: Heinemann Educational Books.
Kelbessa, W. (2005). The Rehabilitation of Indigenous Environmental Ethics in Africa. *Diogenes*, 52, 17–34.
McCullough, M. (2002). Savoring Life, Past and Present: Explaining What Hope and Gratitude Share in Common. *Psychological Inquiry*, 13, 302–304.
Murove, M. (2004). An African Commitment to Ecological Conservation: The Shona Concepts of Ukama and Ubuntu. *Mankind Quarterly*, XLV, 195–215.
O'Neill, J. (1993). Future Generations: Present Harms. *Philosophy*, 68, 35–51.
Opoku, K. (1993). African Traditional Religion: An Enduring Heritage. In J. Olupona and S. Nyang (Eds.), *Religious Plurality in Africa*. Berlin: Mouton de Gruyter.
Oruka, H., and Juma, C. (1994). Ecophilosophy and Parental Earth Ethics. In H. Oruka (Ed.), *Philosophy, Humanity and Ecology*. Nairobi: ACTS Press.
Partridge, E. (2003). Future Generations. In D. VanDeVeer and C. Pierce (Eds.), *The Environmental Ethics and Policy Book*. Belmont, CA: Wadsworth.
Rolston, H. (1975). Is There an Ecological Ethic? *Ethics*, 85, 93–109.
Rolston, H. (2003). Why Species Matter. In D. VanDeVeer and C. Pierce (Eds.), *The Environmental Ethics and Policy Book* (3rd ed.). Belmont, CA: Wadsworth.
Routley, R. (1973). Is There a Need for a New, an Environmental Ethic? *Proceedings of the 15th World Congress of Philosophy*, Sophia Press, Sophia.
Russouw, L. (2003). Why Do Species Matter? In D. VaDeVeer and C. Pierce (Eds.), *The Environmental Ethics and Policy Book* (3rd ed.). Belmont, CA: Wadsworth.
Sindima, H. (1990). Community of Life. In C. Birch et al. (Ed.), *Liberating Life*. Maryknoll: Orbis.
Singer, P. (1975). *Animal Liberation*. New York: Random House.
Tangwa, G. (2004). Some African Reflections on Biomedical and Environmental Ethics. In K. Wiredu (Ed.), *A Companion to African Philosophy*. Malden, MA: Blackwell.
Tooley, M. (1976). Abortion and Infanticide. In L. Beauchamp and T. Walters (Eds.), *Moral Problems in Medicine*. Encino: Dickenson.

VanDeVeer, D., and Pierce, C. (2003). Preserving Biodiversity: Preview. In D. VanDeVeer and C. Pierce (Eds.), *The Environmental Ethics and Policy Book*. Belmont, CA: Wadsworth Publishing.

White, L. J. (1967). The Historical Roots of Our Ecological Crisis. *Science*, 155, 1203–1207.

Wiredu, K. (1994). Philosophy, Humankind and the Environment. In H. Oruka (Ed.), *Philosophy, Humanity and Ecology*. Nairobi: ACTS Press.

5 Transformation of urban space in South-West Nigeria, 2011 to present

Ethical issues in development and aesthetics

Olatunji A. Oyeshile

Introduction

The transformation of urban space to make it aesthetically appealing constitutes a veritable basis for measuring the performance of democratically elected government in South-West Nigeria, especially from 2011 to the present. Concomitant to the renewal of urban space is the displacement of people due to the destruction of living and business structures seen to constitute obstacles in the quest for urban renewal by the government. The displacement of people and structures generates anger, anxiety, tension and hunger, thereby diminishing the popularity of government against the presupposed opening up of urban space. We argue that development is meaningful only when it is not only about aesthetics but also about people who are the ultimate beneficiary. The chapter recommends that states in Nigeria should balance urban renewal with the survival of the people. A gradualist approach that provides a viable alternative source of livelihood is preferred to radical transformation of urban space, given the poverty level of the people, if sustainable development is to be realised. Ethical values must be considered in the quest for urban renewal. Observational, critical and constructive methods are employed in our study. It is concluded that ethical issues should form the basis for any form of development.

One of the most fundamental attributes that distinguish the human animal from the rest of the animals and other living things is the ability to change his/her environment to suit his/her changing circumstances (Oyeshile, 2009, p. 28). This fact underscores the importance of the environment in various dimensions. It can then be argued that the environment provides a place for man from which he can create development and subsequently human happiness. The objective of this chapter, therefore, is to critically interrogate the phenomenon of the transformation of urban space in South-West Nigeria, especially from 2011 to the present, with a view to bringing into focus the ethical issues involved in development and aesthetics due to the transformation of the urban space in the geographical location identified.

The methodology of this chapter is made up of observational, critical and constructive approaches. Just like the natural scientist and naïve participant in the drama of life, we assess the environmental developments, especially concerning the urban space in the southwestern states of Lagos, Oyo, Ogun, Ekiti and Osun States. This serves as the basis of our critical comments on the trend of development in the areas concerned. We noted the merits and demerits of such transformation and then engaged in a constructive approach towards making the transformation viable and sustainable by identifying some key ethical elements that are germane to any development whatsoever.

The Nigerian urban space

The transformation of the Nigerian urban space has been as old as the Nigerian state. However, the 1980s gave more prominence to the issue, especially with military regime of Buhari and Idiagbon (Dec.31, 1983 to August 27, 1985). This was when the War against Indiscipline (WAI) and Monthly Environmental Sanitation Exercise were introduced. Prior to this time, one could aver that certain machineries put in place to take care of the environment, such as sanitary officers popularly called *wole-wole* (checkers of houses), were becoming very ineffective. In fact, one could visualise a palpable journey from squalor to anarchy and to urban chaos.

It should be recognised that environmental health and sustenance play a very crucial role in the alleviation of poverty. For instance, a healthy environment will contribute towards an active, strong and healthy labour force and this means more income. Furthermore, a well-planned environment will reduce poverty because the necessary living infrastructural facilities will be readily available. Hence, the transformation of the environment is both an end in itself (aesthetic appeal) and a means to an end in terms of reduction of poverty and provision of a quality lifestyle. It should also be noted that poverty has always been a major factor determining our human development index.

The fact cannot be undermined that there is a disparity or disconnect between official economic indices indicating economic growth and welfare improvements in the life of much of the Nigerian population. According to the Nigeria Economic Report:

> Poverty reduction and job creation have not kept pace with population growth, implying social distress for an increasing number of Nigerians. Progress towards the fulfilment of many of the Millennium Development Goals has been slow, and the country ranked 153 out of 186 countries in the 2013 United Nations Human Development Index.
>
> (Nigeria Economic Report, 2013, p. 2)

From the above, it is very obvious that poverty is still largely with us, and this points up the need to assess the government's policies, political and otherwise, towards addressing this problem. The urban transformation programme, especially that of the South-West governments of Nigeria, is one such pivotal phenomenon.

Still focusing on the Nigerian urban space, chaos, anarchy and instability that have contributed to poverty are in various dimensions. For instance, the Town Planning or Urban Planning Authorities cannot be said to have discharged their duties to the satisfaction of the public. The question that readily comes to mind is this: if the town planning authority exists to approve all forms of building, private, public and commercial, why are there so many anomalies in our urban space?

Many have responded to this question by citing political factors and corruption as the basis for the ineffectiveness of the Urban Planning Authorities. For instance, it is surprising to see petrol stations sited in residential areas. This practice poses a lot of danger to the people. Yet these projects are approved by the authority even when a red 'x' has been inscribed on such buildings. Again, several houses that do not conform to environmental rules are marked 'x' only for the 'x' mark to be cleaned off when certain exchanges have been made. The problem and chaos are legion. Lastly, the cultural constraints of environmental planning are very instructive. Certain buildings that are taken to be cultural expressions and monuments that constitute environmental constraints are spared as sacred artefacts in spite of their being obstacles to urban transformation. One can then, following from the above, appreciate some of the fundamental issues that define urban renewal or transformation in present Nigeria.

A lot of scholars have also suggested ways in which some of these problems can be surmounted. Many of the suggestions, as plausible as they are, have been more theoretical and academic without taking cognizance of the practical issues involved in the issue of environment. For instance, Dimuna and Omatsone (2010) call for "Regeneration in the Nigerian Urban Built Environment" and Akiode (2012) calls for a proactive integration of theory and practice in the quest for environmental preservation. Again, Olajuyigbe and Rotowa (2011) call for the optimisation of physical planning in the developing countries, while very recently Olufemi and Omirin (2015) stress the need for imagination and utopian thinking in planning cities in the future. The central argument of Olufemi and Omirin can be put in their own words thus:

> The face of the city is changing and cities have always been a juxtaposition of good and evil, entertainment and edutainment, order and disorder, ugly and beautiful, pleasure and pain, productive and unproductive, consumption and waste. Imagining the future through the renewed lens of utopian thinking will make for peaceful and harmonious cities . . . the planning profession should begin to engage, develop tools and imagined principles to accommodate or manage the emerging multiple utopias such as sustainable utopia (green or eco-friendly, carbon-neutral city) or technological utopias (controversial genetically modified foods/organism – GMO, GIS/CAD/GPS, 3D, Internet, Twitter etc.) or sitopia – foods place, or cyberspace or lunar utopia – in the places and spaces where people live, work play and learn.
>
> (Olufemi and Omirin, 2015, p. 34)

Our response to the above is that the utopian thinking concerning the planning of cities in Nigeria is often hampered by such factors as poverty, political will and politics.

South-West Nigeria: urban transformation, 2011 to present

The South-West of Nigeria constitutes one of the most politically enlightened, economically and educational developed regions of Nigeria. The transformation of the urban space in South-West Nigeria, especially in Oyo, Lagos, Ogun, Osun and Ekiti (under the All Progressive Congress) states, to make it aesthetically appealing constitutes a veritable basis for measuring the performance of democratically elected governments in the states mentioned above.

It may be asked, what constitutes this urban transformation? To every discerning observer, the transformation has been mainly in the areas of opening up urban roads and erecting public commercial infrastructures such as shopping malls. Prior to the coming of the All Progressive government in these states, the urban spaces have not only been congested, they have also been dirty and chaotic. In fact, traders display their wares on major roads competing, more often than not, with vehicles for space. Again, traders display their wares under high-tension electrical installations, while markets have no access roads in case of emergencies, such as fire incidents. In all these, many lives have been lost and properties of unquantifiable cost have been destroyed, leaving the populace poorer and unhappy. The bid of the populace to survive economically has been a trade-off for quality of life and life sustenance.

From 2011 to the present, there is no doubt that the urban areas have been transformed for better. In some cases, new markets have been built to accommodate displaced traders in many of these states. But it is not clear whether most of these markets have conveniences that would make their relocation worthwhile. Again, lives lost as a result of trading on the road or under high-tension electrical installations have been reduced. Furthermore, the environments are now more beautiful. There is a sense of aesthetic appeal in many of these states. Of course, this factor is often over-exaggerated to a boastful level when it is claimed that construction of overhead bridges has improved the quality of life of the people.

What are some of the problems associated with urban renewal or urban transformation? The problems are many. Some of them include destruction of houses, wares, shops and other means of livelihood. In some instances, it has resulted in life-threatening trauma for some people who could not bear the shock of the loss of their means of livelihood. Some states in the South-West, especially Oyo State, have set up a judicial body of enquiry saddled with the task of paying compensation to people whose properties were unlawfully demolished or who had the prior government's approval before such buildings were erected in the first instance.

Environment and development

Every human society is constantly preoccupied with the urge and task of development. Even the so-called 'developed Western' societies daily strive to achieve development in one area of human endeavour or the other. Development, then, in this context, is to be understood in terms of a viable standard of living, hope for the future, trust in leadership and a sustained social order for the actualisation of individuals' rational life. According to Bamiro, "development definitely has the pride of place as one of the most engaging issues of our time" (Bamiro, 2008, p. iv).

At both theoretical and practical levels, there have been divergent views as to what constitutes development and how it can be achieved, but one can claim that there is a convergence that development is not guaranteed by simple infrastructure development or a rise in a nation's Gross Domestic Product (GDP) or natural resource endowment. Rather, there is a strong advocacy for sustainable development "seen as a multifaceted improvement in the lives of the people and increase in their capability and enablement to define their own problem and find their own solutions in harmony with their environment" (Bamiro, 2008, p. iv).

The above analysis shows that development is not predicated on a monolithic perspective, as it was at a given time in history. For instance, at a point it was assumed that development was simply economic development construed as the rise in a country's Gross Domestic Product (GDP). While it is true that the analysis of development is now pursued from a multifaceted perspective, it is still instructive to note that the analysis of development from a certain unitary perspective is not only useful, but it cannot be ruled out on some occasions, as this determines the focus of development at particular periods.

Given our trend of argument, therefore, development could be seen as "the pursuit of improvement in Gross Domestic Product"; "the pursuit of improvement of human relationships through the promotion of human values such as freedom, justice, equality and co-operation" (Igbafen, 2003, pp. 1–2). As a developing society, we cannot avoid approaching development from specific areas because this is the only way we can ensure a holistic development. In this regard, Akin Mabogunje (1980), as an urban geographer, construes development as a strategy of spatial reorganisation which is crucial for the whole process of political mobilisation and of central state control over planning of productive forces, not necessarily concentrated at just a spot.

Let us note at this juncture that the ultimate goal of development is human well-being and it is pursued from various perspectives, such as human development, environmental development and spiritual development. Under human development we have personal development, which comprises but is not limited to education, health and mental development. With human development we also have economic development, political development, social development, moral development and cultural development. Environmental development involves the sustenance and improvement of the environment, while spiritual development involves the quest by man for transcendental values as seen in their religious activities.

It should be noted that development has material and non-material components. The two components cannot be pursued in a mutually isolated form, as the improvement in one necessarily affects the other. A society is developed, therefore, when it is able to bring out the latent or potential qualities in its individuals and environment to a more advanced and highly organised state which will make living more pleasant. The question can be raised: why is there a demarcation between development and human development? There is demarcation because although the notion of development has man and his well-being in focus, it nevertheless transcends man to include the environment. Within the environment we have other physical components and living organisms whose preservation and improvement will contribute to human well-being.

It is then easy for us to see that, while development is multifaceted and holistic, its attainment must be engineered through a piecemeal method. This then takes us back to the definition of development. Mabogunje's view that development constitutes a strategy of spatial reorganisation (Mabogunje, 1980) is very apt. Having set our agenda on what development is, let us then examine the place of environment in development.

A major problem facing the global community today has to do with the protection of the environment as a result of the depletion of the ozone layer and the resultant global warming, which poses a great challenge to the human species.

While we are concerned with the problem of global warming, there are other fundamental issues about the environment that we need to come to terms with in Africa. Some of the issues concern the maintenance and provision of infrastructure amenities in our environment, beautification of the environment and, by extension, sustaining the inhabitants of the environment through the provision of life-sustaining measures.

The development of the environment leads directly and indirectly to the sustenance of individuals. It is interesting to note that some scholars have pointed out seeming conflicts between developmentalism and environmentalism. Environmentalism, according to Olu-Owolabi, refers to the philosophy that "celebrates and propagates the value of environmental sustainability" (Olu-Owolabi, 2009, p. 112). The philosophy simply stresses the need to sustain the environment because the natural resources which humans exploit and explore can be exhausted if caution is not exercised.

Developmentalism, on the other hand, has to do with the desire to realise the ideal of development by any means possible (Olu-Owolabi, 2009, p. 112). Developmentalism is predicated on the wild pursuit of material transformation for human benefit with little or no regard for the natural environment from which the natural resources are taken.

The developmentalism thesis is anchored in the belief that "the purpose of natural resources is simply to satisfy human needs and interest and hence there is absolutely no reason for humans to treat natural resources with any sense of caution" (Olu-Owolabi, 2009, p. 112). Given this conception of developmentalism, "development is therefore regarded as the ultimate value, the ultimate end, while all natural resources are indeed only means for the realisation of this end" (Olu-Owolabi, 2009, p. 112).

One can discern that the two positions must be reconciled to pave the way for human progress since man has to exploit natural resources and at the same time protect the environment from which the natural resources come. It is based on this that the World Commission on the Environment (1987) defines sustainable development as: The use of the environment for development in such a way that the needs of the present generation shall be met and that the future generation shall be guaranteed.

It must be noted that some philosophers and environmentalists have placed the issue of the environment at the core of African development. Ekwalo attempts a commanding analysis of the environment in his work *Western Philosophical and Historical Attitudes to Environmental concerns: A Critical Exposition* (2012). He shows the importance of environmental studies in our world today and in Africa specifically but this is beyond the scope of the present work.

Ethical issues arising from the transformation of urban space in South-West Nigeria

In this section, we raise some ethical issues, or what otherwise might be called moral issues in the transformation of the urban space in South-West Nigeria. This is crucial because the phenomenon of urban transformation itself is a means towards an end, which is human development and, by extension, human happiness. In doing this, let us start by taking a look at the notion of moral values.

The notion of moral values

In any human society, social interaction would be not only difficult but impossible if people did not have a sense of right and wrong. This sense of right and wrong is important if the individual is to have a good life. In striving towards his own welfare or self-interest, the individual is also aware that the welfare of others is as important as his and therefore that he must act in a way that will ensure not only his own well-being but the well-being of the society in which he finds himself either by accident or by a deliberate act.

Since we live together, a sense of right and wrong, good and bad, is germane for the sustenance of society. An action is thus moral if the consequences, either positive or negative, will affect the well-being of others, even though actions that threaten the self or agent are also considered immoral. It then bears pointing out, following Pettman, that:

> Morals are social products, held out of habit and self-interest to be sure, but motivated also by emphatic urges like love, and a sense of what is right and due. Human beings learn to prefer particular ways of relating to each other and standards by which to assess their relationships and to endorse or censure what appear to be sociable or anti-social acts, as part of the general educative process whereby they assimilate expectations of how the world is meant to be.
>
> (Pettman, 1979, p. 19)

Moral values then encompass those ingredients that would make a moral situation achievable. They would include not only such values that would make self-preservation possible, but also such values that would ensure the preservation of others by enlightened self-interest on the part of various individuals. They include honesty, avoidance of greed, avoidance of lies, trust, keeping of promises when we are not presented with a dilemma, altruism, that is, placing others first, love and obedience. Although the list is not exhaustive and may vary from culture to culture, the nexus of all these values lies in the fact that they help to ensure human well-being.

Although moral values may vary from culture to culture, there is a sense in which we can talk about the universality of morality. The primary consideration for this is the rationality of man. In this regard, when we say that man is a rational being, we are implying, according to Sogolo, that there are certain things shared in common by mankind and if these traits, such as love, self-reflection, honesty, sympathy and so forth, are absent in any human group, there will be doubt as to whether that group is human at all (Sogolo, 1993, p. xv).

Moral values and contemporary African society

Our use of the phrase contemporary African society, while not implying that there is a uniformity among various African societies concerning their approach and reaction to the issues of knowledge, religion, socio-political organisation and reality as a whole, is appropriate here in the sense of the universality of moral values and for the fact that most African societies are deeply enmeshed in similar moral disequilibrium that has made purposeful leadership, good governance and development daunting tasks. Let us note also that social, communal and ethical values are intertwined in such a way that one cannot talk about one without the other. This simply underscores the fact that moral values have social origins.

Bewaji poignantly describes social basis of morality thus:

> It is the social milieu in which competition for the scarce resources of the environment takes place. But it is not only the resources of the environmental that are scarce. The human resources of love, patronage, recognition, compassion, companionship, etc., are also scarce and require deliberate efforts in both their generation and equitable distribution. Here lies the crux of the moral responsibility of society to its members and to itself. And this fact is represented in numerous ideas in African thought.
>
> (Bewaji, 2004, p. 397)

The extensive passage from Bewaji has a lot of implication for contemporary Africa in the sense that we in Africa not only are experiencing the dearth of material resources needed for development, but also lack the moral resources to manage the little resources we have due to what Olusegun Oladipo has described as "moral dislocation" (Oladipo, 1998). The present situation is sharply in contrast to what prevailed in traditional African societies, in which brotherly concepts such as *Ubuntu* flourished. It is true that social order exists to provide the multifarious needs of man which are necessary for a fulfilled happy life conducive for development. It is also true that social order is a condition in which society is organised to effectively provide for the needs of man. Granted that this is the case, how do we in Africa ensure the desired social order? This is where morality comes in. Interestingly, morality leads to the sustenance of social order, and social order can enhance not only the internalisation of morality, but also its application.

Let us note again that this moral order is predicated on certain values described variously as communal, social and human values. According to Sogolo, human values can be taken to be essentially natural and universal simply because human nature itself is universal. He puts human values thus:

> Human values are intrinsically designed to fulfil man's goal of individual and collective survival. Therefore, the human tendencies to pursue virtue such as peace, care, justice, truthfulness etc are naturally embraced to ensure the well-being and survival of humankind.
>
> (Sogolo, 2004)

The question again is this: What has made the internalisation or application of moral or human values difficult in Africa? The answer can strictly be situated within our socio-political and economic sphere. The state of moral values in Africa has been exposed to unfavourable economic pressure, political competition with its attendant vices such as corruption, violence, ethnic cleansing, injustice in the allocation of natural resources. It would take an angel to be moral in a situation in which survival in the political sphere is not guaranteed and in which the political space bordering on justice, freedom and equality continues to recede on a daily basis.

Having thus examined the notion of moral values and their importance in our social organisation, or what existentialists would call "our-being-in-the world," it is now instructive to assess some ethical issues in urban transformation in South-West Nigeria.

The first issue that comes up is that of the nexus between aesthetics and hunger. Do we pursue aesthetics while we make the people hungrier, especially in the short run? It will be immoral to sacrifice the people's happiness for aesthetics all in the name of beautification, no matter what benefits will accrue to people in the long run. A critical assessment of the transformation of the urban space in the South-West shows that many of the governments did not think of alternative sources of livelihood for the people. At the end of the day, many families whose living and business structures were demolished found it difficult to survive. This automatically raises the question of whether "aesthetics is made for man or man for aesthetics?"

The second moral issue, which is related to the first, is the means-end relationship. In ethics, any action that sacrifices the human person for a so-called desirable end is immoral. In the process of urban transformation in South-West, many human agents were negatively affected either with regard to their comfort or means of livelihood. Our position here should not be mistaken to mean that no sacrifice should be made in our development stride. What is being emphasised here is that the governments concerned should have made alternative arrangements for the people before demolishing their properties. These alternatives could be in the form of makeshift temporary houses or markets. What we found out is that alternative markets were built after the so-called road shops and houses were demolished. In some cases, the monetary compensation does not get to the people on time.

In engendering urban transformation, the concept of environmental ethics should be brought into focus. Furthermore, such ethical theories as Utilitarianism and Situation Ethics should also be given considerable attention. For instance, utilitarian ethical theory advocates that an action is right if it produces the greatest happiness for the greatest number of people and wrong if it does not promote the greatest happiness of the greatest number of people. In a similar vein, situation ethics will deem an action to be right if it is carried out based on the existing situation. In this vein, certain variables must be taken into consideration, such as the poverty level of the people, their survival and happiness. It is the case then that the transformation of urban space in South-West of Nigeria is very welcome and desirable, but the method of doing it violates certain ethical principles concerning human happiness.

Conclusion

Towards a holistic, human-centred urban transformation in South-West Nigeria

Urban transformation is part of human development. If it is well articulated and implemented, it reduces poverty and enhances development and human happiness. Our recommendation is based strongly on the fact that 'development is made for humans and not humans for development.' If this is true, any form of urban development must be all-encompassing, taking into consideration all facets of human life.

For instance, if, by our developmental actions, the human agent is decimated, then when the development is completed, who will be the beneficiary of the development? Certainly, no government will subscribe to this form of development. Therefore, it is suggested that any government should consider primarily the people's survival in the short run and long run in the quest for urban transformation.

Furthermore, the transformation of the urban space will be a mirage given the present level of rural-urban migration, especially in South-West, Nigeria. The major way in which we can have sustainable urban transformation is for governments to concentrate more in developing the rural areas through the provision of necessary infrastructure. The neglect of the rural areas, which are supposed to be the major economic foundation for modern states, has led to uncontrollable drift to urban areas. Consequently, there is much pressure on the facilities in urban areas. These facilities subsequently break down, taking us back to our crude living.

Lastly, government should exhibit enough political will to ensure that those saddled with the management of the environment carry out their duties. Unnecessary political interference, party politics and corruption have been the bane of environmental development in Nigeria. While one must commend the governments of South-West Nigeria for taking the bold step in transforming the urban space, they should also pay serious attention to ethical issues, such as human happiness, arising from such efforts.

References

Akiode, O. M. (2012). Proactive Integration of Theory and Practice in the Quest for Environmental Preservation: A Philosophical Appraisal. *Journal of African Environmental Ethics and Values*, 3, 71–89.

Ayotunde, J. and Bewaji, I. (2004). Ethics and Morality in Yoruba Culture. In K. Wiredu (Ed.), A Companion to African Philosophy. Malden, MA: Blackwell Pub. Ltd. pp. 396–406.

Bamiro, O. A. (2008). Foreword. In L. Poopola et al. (Eds.), *Contemporary Issues in Development: Lesson for, and Challenges to Nigeria*, 1. Ibadan: The Postgraduate School, University of Ibadan.

Dimuna, K. O., and Omatsone, M. E. O. (2010). Regeneration in the Nigerian Urban Built Environment. *Journal of Human Ecology*, 29(2), 141–149.

Ekwalo, C. J. (2012). *Western Philosophical and Historical Attitudes to Environmental Concern: A Critical Exposition*. Lagos: University of Lagos, Department of Philosophy.

Igbafen, M. L. (2003). *Basic Issues and Theories in Philosophy of Development*. Benin City: Lucosen Publishing House.

Mabogunje, A. (1980). *The Development Process: A Spatial Perspective*. London: Hutchinson & Co Publishers.

Nigeria Economic Report. (2013, May). No. 1.

Oladipo, O. (1998). "Modernisation and the Search for Community in Africa: Crisis and Conditions of Change" O. Oladipo (Ed.), Remaking Africa: Challenges of the Twenty-first Century. Ibadan: Hope Publications, pp. 106–123.

Olajuyigbe, A. E., and Rotowa, O. O. (2011). Optimizing Physical Planning in the Developing Countries – a Case Study of Ondo State, Nigeria. *Journal of Sustainable Development*, 4(4), 202–209.

Olufemi, O., and Omirin, O. (2015). Utopia Thinking: Imagining the Future of City Planning. In *Book of Abstracts*, 2nd Faculty of Arts, University of Ibadan Biennial International Conference. Theme: Imagining Order/Other – the Humanities and Global Peace and Security, 18–22 May.

Olu-Owolabi, K. A. (2009). A Critical Reconstruction of Anyiam Osigwe's Environmentalism. In O. Oladipo and A. B. Ekanola (Eds.), *The Development Philosophy of E.O. Osigwe Anyiam-Osigwe* (Vol. 1, pp. 108–121). Ibadan: Hope Publications.

Oyeshile, O. A. (2009). Mind-Change on Environmental Development and Street Rebranding: A New Paradigm for Sustainable Development. *Mushin Maiden Summit on Street Renaming, Renewal & Rebranding*, 28–44.

Pettman, R. (1979). Moral Claims in World Politics. In R. Pettman (Ed.), *Moral Claims in World Affairs*. London: Croom Helm.

Sogolo, G. (1993). *Foundations of African Philosophy: A Definitive Analysis of Conceptual Issues in African Thought*. Ibadan: Ibadan University Press.

Sogolo, G. (2004). Philosophy, Human Values and Social Order. A Keynote Address Presented at the *30th Anniversary Celebration of the Department of Philosophy*, University of Ibadan, September 16.

World Commission on the Environment. (1987). Report of the World Commission on Environment and Development: Our Common Future. Oslo: WCED. Retrieved from http://www.un-documents.net/our-common-future.pdf

6 Animal rights vs. animal care ethics

Interrogating the relationship to non-human animals in Yorùbá culture

Ebunoluwa Olufemi Oduwole and Ademola Kazeem Fayemi

Introduction

Recent developments in the cruelty, suffering, starvation and means-end treatment of non-human animals have led to renewed interests, areas and theoretical leanings in environmental ethics in Western intellectual discourse. The non-human animal liberation and rights movements and the animal welfare advocacy group have introduced prominent views dominating the prescriptive debate on what non-human animals are owed. In many African cultural and intellectual spaces, such attraction and advocacy of animal liberation and rights are very low in priority. Indeed, until recently, ethical theorising about the moral status of non-human animals in the African worldview received little attention.[1] Many African scholars are much busier ruminating on how to change the worsening human conditions in the continent rather than caring about the condition of the non-human animals they co-habit the environment with. While such human-centric devotion is not in principle wrong, there is also the need to interrogate and reflect on the nature of the relationship existing between humans and non-human animals in Africa. This chapter aims at examining the nature of this relationship by situating the discourse within the context of the traditional Yorùbá-African culture.[2]

The fundamental questions are: What are the defining moral principles underlying the attitudes and perceptions of the traditional Yorùbá towards non-human animals in their cultural milieu? Is there a correlation between the ontological perception of non-human animals and the ethics of human relationship with them in Yorùbá culture? Does the culture reflect the position of animal liberation, animal rights or animal care or an entirely different perspective from the known Western taxonomies?

This chapter shall attempt to explore answers to the above raised questions by investigating the nature of the values that underpin the indigenous Yorùbá relationship with and understanding of animals. In view of the absence of documented history of non-human animals and the human relationship with them among the Yorùbá, the people's oral tradition is one cogent source of discovering their attitudes and practices. While there are other ways of validating knowledge in contemporary African cultures, this chapter shall explore oral traditions, specifically, in order to know what kinds of values underpinned the traditional Yorùbá perception of non-human animals. This is without prejudice to the possible limitations that oral traditions as an epistemic source may involve.

The justification for the reliance on oral tradition is that, as a general principle, actions are predominantly expressed and codified in speech; as a general principle, the Yorùbá are not an exemption to this. For them, "to be in traditional society is to be in oral tradition, and what is not codified in oral tradition, is most likely unknowable and unreal" (Fayemi, 2010). To discover ideas, their existence and epistemic status in the Yorùbá worldview, it is instructive to examine the oral tradition, hermeneutically. This is particularly important because such oral traditions are deduced from terse formulations on reflections about close observation of life, life forms and their characteristics and habits, natural phenomena and the environment at large. Proverb is an instance of the Yorùbá oral tradition. African environmental ethics, and animal ethics in particular, is an understudied literature using the hermeneutic approach.

In engaging in a hermeneutical analysis of the traditional Yorùbá perception of and relation with non-human animals, this chapter aims primarily at discovering how best to theoretically frame such attitudes and practices within the bourgeoning global discourse on animal ethics. Given the dearth of normative theorisation on human–non-human animal relations in Yorùbá intellectual culture, this hermeneutic concern is not motivated towards claims of uniqueness, as many other traditional societies in Africa and beyond may also have such a theoretical account. Rather, the objective is to discover the Yorùbá normative conception of human–non-human animal relations, and to show that there are other perspectives in addition to the dominant animal liberation and animal rights positions, which are illuminating for a cross-cultural understanding of human-non-human animal relations. Adopting this hermeneutic approach will also be helpful for understanding the extent to which traditional insights could be useful in effectively managing current challenges of non-human animals across regional boundaries.

The chapter proceeds with a critical discussion of the debate in mainstream Western animal ethics. Following this, the insights from indigenous Yorùbá animal care are provided, with some comparative notes.

On animal ethics

Is there an ethical obligation towards lower or non-human animals? Peter Singer is one prominent scholar who answered in the affirmative. The general unequal treatment of animals is of concern to him. In his *Animal Liberation*, Singer argues that while "all animals are equal," human experiences have continued to show the cruelty, suffering and means-end treatment of non-human animals. Premising his defence of animal liberation and animal equality on Jeremy Bentham's utilitarian principle, Singer argues that human and non-human animals are to be grouped "exactly in the same way." As a matter of biological fact, both experience pleasure and pain. Sentience, which means the capacity to experience happiness and suffering, is a common denominator of human and non-human animals. However, grouping both human and non-human animals together does not result, in practical terms, in the equal treatment of animals. This is Singer's lamentation and he gave grounds for why the unequal treatment deserves attention.

On the basis of the principle of utility, Singer maintains that "if it is wrong to cause human suffering to achieve a good that does not outweigh that suffering, then "it will be wrong to cause nonhuman suffering to achieve such a good" (Camp, Olen and Barry, 2015, p. 445). Singer urges that we should maximise the happiness of animals and minimise their suffering. He argues that, as a matter of valid logical consistency, what makes racism unjust in human-to-human relations is not fundamentally different from why it is wrong to oppress non-human animals by using them for experimentation, sports, entertainment and food, among other uses. Singer, therefore, argues that non-human animals and humans are equal and that this "equality does not depend on intelligence, moral capacity and physical strength. Equality is a moral ideal" (Singer, 1989, p. 151). He further notes that "the basic principle of equality does not require equal or identical treatment. Equal consideration for different beings may lead to different treatment and different rights" (Singer, 1989, p. 149).

Singer thinks that sentience is a more fundamental justification for treating all animals equally. If one were to accept the supposition that features such as rationality, self-consciousness, will and intelligence are conditions for human moral rights, then, Singer argues, the inference will be that "we have the right to perform painful experiments on retarded humans for trivial reasons; similarly it would follow that we had the right to rear and kill these humans for food" (Singer, 1989, p. 158). The main reason for this inference is that the mentally retarded lack the capacities popularly thought of as necessitating moral rights. The point Singer (1989, p. 154) is driving at is that "the suffering that humans inflict on animals while alive is perhaps an indication of speciecism," and that speciecism is wrong. Speciecism is the prejudiced discriminatory attitude rooted in differences in species as a basis for supporting actions and interests of one's own species' members.

While one agrees with Singer's claim on needless cruelty to animals, some critics have argued that "we have no obligation to promote the good of nonhuman animals at humankind's own expense" (Camp, Olen and Barry, 2015, p. 447). Singer's position on the equal treatment of non-human animals with humans entails the obligation of non-consumption and non-experimentation of animals even if doing so increases the well-being of humans. Singer's position has been argued to be an absurd interference with nature and against normal moral intuition. As Callicott et al. (2011, p. 140) noted, "extending rights to life into the natural world (the animal world) would wreak havoc: it would mean protecting prey from predators."

Against Singer's animal liberation framework and defense of non-human animals in moral consideration, Tom Regan argues that it is a wrong logical entailment to suppose that all actions causing suffering entail cruelty to animals. He argues that contrary to speciecism, the most important principle of action for moral consideration is rights-based. This is because respect for persons requires that we do "not treat other humans in certain ways, regardless [of their lack] of utility" (Camp, Olen and Barry, 2015, p. 447). Regan, therefore, modifies Singer's view by arguing that humans should have respect for at least some non-human animals that are experiencing-subjects-of-life, just as humans are. Animals that are experiencing-subjects-of-life are conscious of their environment, have desires and feelings, memories and preferences, just as they also have a sense of self-identity and a vision of the future. Adult mammals do meet these criteria; hence, their lives are of independent value. It will therefore be morally wrong to use or treat them as a means to our own end.

Kenneth Goodpaster (1978) rejected both Singer's equality/suffering criterion and Regan's subject-of-a-life criterion as bases for non-human animals having rights. In both, the value of plants and their moral rights are not preserved. Rather, it makes more sense to use 'life,' which all biological entities share in common, as the criterion of moral consideration. In his words, "nothing short of the condition of being alive seems . . . to be a plausible and non-arbitrary criterion" (Callicott et al., 2011, p. 140). While all biological life deserves moral consideration, Goodpaster is quick to add that not all life has the same moral significance as there should be a limit to the respect for non-humans. But this distinction between moral consideration and moral significance raises the question of how to prioritise and measure the moral significance of species. It also raises the validity of the analogy between humans and animals, as only humans are governed by moral laws. Critically, Goodpaster fails to provide any criterion for knowing and demarcating the morally significant beings. Besides, his account of moral rights, just as those of Singer and Regan, undermines the natural environment such as air, landscapes and rivers, among others, which presumably lack life in Western intellectual culture.

Paul Taylor's (1992) view is an attempt to remedy the above deficiency in thought when he came up with non-human rights and an attitude of respect for nature in his proposal of legal protection for all species. For Taylor:

> It is the good (well-being, welfare) of individual organisms, considered as entities having inherent worth that determines our moral relations with the Earth's wild communities of life . . . [Hence], humans are morally bound to protect or promote their good for their sake. . . . Each individual organism is conceived of as a teleological centre of life pursuing its own good in its own way . . . the claim that humans by their very nature are superior to other species is a groundless claim [to be jettisoned].
>
> (Taylor, 1992, pp. 95–96)

For Taylor, every living organism has inherent worth and thus deserves equal moral consideration. The fundamental shortcomings of Taylor's biocentric view is that not only does he erroneously ignore the "differences of moral significance among the bearers of moral standing" (Attfield, 2009, p. 99), but he also extends biocentricism to an absurd extreme (Engel, 2008). "Even if animals had and were afforded intrinsic value, this in itself does not guarantee that human (basic) interests would not trump animal interests in moral evaluation" (Ahteensuu, 2016, p. 290).

Following the foregoing brief overview of the main currents in contemporary Western ethical discourse on the values that ought to underpin human relations to animals, it is important to further enrich the moral deliberations on human-animal relationship by a consideration of a non-Western perspective. In this regard, the next section shall discuss the traditional Yorùbá view, which has received little attention in African animal ethics discourse.

African worldview and animal care: the Yorùbá example

Understanding people's religious and metaphysical orientation is important in cognising their attitude to non-human animals. The Yoruba believe in a cosmological world created by *Olodumare* (Supreme Being) in conjunction with his divinities. For them, the environment consists of two parts: the visible and the invisible part. These two parts involve a cosmological hierarchy where the invisible is the higher plane to the visible part. While the visible part non-hierarchically consists of plants, trees, mountains, rivers, human beings and non-human animals, the invisible part according to Odera Oruka consists of "the dead, the unborn, the hidden stars, planets and the spirits, whether of human beings or other creatures" (Oruka, 1997, p. 252). Some of these natural and visible features also serve as abode for a category of the divinities (Awolalu and Dopamu, 1979, p. 73). However, the division between the visible and invisible planes is not water-tight; there are interrelations as the visible aspects of nature are sometimes personifications of invisible forces.

Whatever people consider as the home of a spirit (be it rivers, hills, mountains, caves, brooks, waterfall, streams, trees, thick forest, etc.) is usually set apart as a sacred place, and, there, people offer worship to a particular spirit (Ogunade, 2005, par. 3). The belief is that the features in nature are not accidental but created and placed there for significant purposes. While human beings are important stakeholders, their existence finds expression in their association with non-human visibles and the non-visibles; without the harmonious association, human beings are believed to be incomplete (Mangena, 2013, p. 30). In view of the religious beliefs and their concomitant rituals, many of these natural artefacts of the environment are revered. The traditional Yorùbá used rituals in correcting some violations of nature's integrity.

There is a religious allegiance to animals as objects of worship and, as such, some animals are venerated and respected. Some animals within the Yoruba traditional worldview are revered as sacred figures of religious worship, as the people tend to venerate the spirits behind some animals. Some of these animals include the yellow palm bird (popularly called *eye oga*), vulture (*igun*) and royal python. Any attempt to fell a tree where the decorous bird (*eye oga*) lays its eggs is to visit doom upon the society (Owoseni and Olatoye, 2014, p. 109). Animals such as these simply become totems and are regarded as sacred, enjoying a privileged place in the Yoruba community (Owoseni and Olatoye, 2014, p. 110). The vulture in Yorùbá land is a sacred bird and should not be used as a burnt offering, game or food. The Yoruba saying confirms this: *a ki pa igun, a ki je igun, a ki fi igun bori* (Adewoye, 2007, p. 54) – "We do not kill the vulture, we do not eat the vulture, we do not use the vulture as sacrifice to the gods to remedy human destiny." This saying warns against any attempt to kill the vulture for food, or use it as sacrifice to the gods.

The Yoruba tradition further encourages the preservation of animals through certain rituals, customs or taboo. The Yorùbá believe that proper sacrifice must be offered to the deities in order to avert calamities and to also propitiate the gods for their offences and mistakes that may incur the wrath of the deities upon them. Deities in Yoruba culture are numerous and each of them is associated with a particular animal, which is sacrificed to this particular deity. For example, Ògún (god of iron) is very fond of dogs, snails, tortoises and rams; Òsun (a goddess of the river called by that name) accepts goats and fowls; Èsù (the Yorùbá trickster deity) is fond of black fowl; Ṣàngó (god of thunder) is fond of ram; Òrúnmìlà (god of wisdom, knowledge and prophesy) is fond of rats and mud-fish; Obàtálá (the Supreme God) is fond of snails; Òsanyìn (the god of herbal medicine) is fond of tortoises; Egúngún (Masquerade) is fond of rams, just to mention a few. In all, fundamental ethical questions can be raised and problematised around such ritualistic practices involving animals: are such practices expressions of cruelty against animals or are they expressions of respect and reverence of the spirits of such animals?

While it is arguable that the "reverence, responsibility and restraint in connection with interaction with sacred [kinds of animals] are based on the fear of ancestral spirits" (Taringa, 2006, p. 213), however, such an ecological attitude is moral only by default. In truth, while the traditional Yoruba care for and protection of animals are based on fear or respect of ancestral spirits than on respect for the [animal in] itself (Taringa, 2006, p. 191), it is paradoxical that such reverence does still take the animal as a moral patient. Respect for animals is religiously inclined, but moral by default, with humans assuming the status of moral agents while animals and things in nature are seen as moral patients to which humans owe certain obligations. "Moral patients are morally entitled to human ethical concern; moral agents have the means (their thought and actions) to provide their concern or care" (Oruka, 1997, p. 251). Being moral patients does not imply the weakness of the non-human visibles (like animals), or the passivity of the invisibles. Indeed, the invisibles "have greater power and knowledge than the visibles" (Oruka, 1997, p. 252). But the relational metaphysics underpinning the Yoruba worldview explains the elasticity between the visibles and the invisibles.

"The metaphysical structure of the Yorùbá conception of the nature of reality falls within the binary complementary conceptual scheme" (Fayemi, 2012, p. 13). Binary complementary metaphysics entails that there is a force permeating everything which explains the connectivity and interdependence of all things, humans, plants and animals and the non-living entities in the environment. Such a metaphysical scheme is grounded in interdependence, inseparable and harmonious relationships of both the visible and the invisibles constituting the environment. This "indissoluble solidarity between humanity and the environment" (Murove,

2004, p. 202) implies sharing certain norms that foster harmonious relationships not only among humans but also between humans and non-human visible components of the environment, as well as the non-visible constitution of the environment. The binary metaphysical idea of togetherness and relationalism between human, non-human animal and inanimate things and components of nature has implications for an African animal ethic.

Within this understanding of reality, there is a religious interpretation of animals. By this, it means that some animals among the Yoruba are granted 'divine' privilege and are revered. For instance, there is a myth surrounding the reverence for the river goddess *Oya* and buffalos, which are exempted from the category of game animals to be hunted. The religious-metaphysical expression is also accounted for in the Yoruba belief in the transmigration of human spirits into the bodies of animals. Though this sort of thought is mysteriously rather than scientifically grounded, it accounts for the Yoruba belief that "possessed" animals are perpetrators of both evil and good deeds.

In Yoruba cultural understanding, there are patterns of both unhealthy and healthy relationships between humans and other animals. While this worldview demonstrates an awareness of the consciousness of animals and their experience of pain and suffering, it conceives of animals as nutritional, a consumable means of promoting human health and satisfaction:

> *Ife ti a fe adie ko denu, ibi ki a paaje lo mo* – "Our love for roosters is not genuine; the point is to make it a delicacy."
>
> (Ajibade, 2006, pp. 168–169)

The above proverb illustrates that the Yoruba reckons that animals are sources of human food, despite human awareness that killing animals causes them pain. The consumption of animals surpasses the purpose of nutrition to include medication/treatments (healing, in Yoruba land), as animals' bodily parts are used alongside other curative ingredients to treat specific ailments (Ajibade, 2006). Indeed, the instrumental use of animals as companions within domestic folds is a rarity in traditional Yoruba society. However, such a practice is now becoming common in contemporary Yoruba societies owing to the cross-cultural influences and infiltrations of Western lifestyles among the peoples.

From the above, there is the tendency to say that the Yorùbá tradition does not display in clear terms the sphere of its ethical viewpoint regarding human-to-animal relationships. Some sayings, proverbs and beliefs appear to be 'equality' oriented, or to represent a welfarist concern for animals.

The Yoruba say: *ise eniyan nise eranko* – "the ways of man/humans are also the way of animals."

Often, such sayings have dual meanings, as both metaphoric and literal expressions in reference to human and non-human situations. The saying above implies, in a way, the idea of semblance as opposed to equality of human and non-human observed patterns of behaviour. So the idea of equal treatment advocated by Singer is not dominant. Also not applicable to the Yoruba is Singer's moral judgmental grouping of 'speciesists.' While speciesism in Singer is a function of discriminatory acts and valuations against other species simply on the grounds of non-membership of one's own species, which for him is morally unjustified, speciesism for the Yoruba, at best, suggests the prejudiced bias involved both in ageism migration from being young to old or vice versa, and in the non-distinctive capacities that confer temporal moral status on humans because of the non-recognition of such capacities in non-human animals.

However, like the animal liberation/rights theorist in the West who extends moral duty to animals on the basis of sentience, the traditional Yoruba also recognised the sentient nature of animals and other species. Proverbs depicting this point are:

i *Ojú ni màlúù n ró, òbe ò dára lórùn* (Olusola, 2006, p. 159). (The knife is not something pleasant on the neck; the cow suffers pain in slaughtering)
ii *Bí ó ti ńdun ọmọ ẹyẹ, bẹ̀ẹ̀ ló ńdun ọmọ èèyàn* (As the young birds are hurt, so do humans experience pain)

These proverbs agree with the sentientist view that other species feel pleasure and pain, just as humans do. The Yorùbá are aware that animals have their own emotions and feelings, and that the pains of animals are morally worthy of care consideration and not equal consideration as advocated by Singer. The proverb below, for instance, challenges the tendency to ignore the pains and sufferings of non-human animals by humans:

Akeyinje ko mope idi n ro adie (Adewoye, 2007, p. 56) – "The person who gluttonously consumes the egg does not know the pains the hen passed through in laying it"

Such proverbs oppose non-humane consideration of animals by condemning the recklessness and lack of empathy of some humans to the reproductive process of non-human animals.

However, one needs to quickly note that, unlike the animal liberation/rights theorist who contends that the only right way to treat animals is to treat them as ends in themselves, and not as a mere means (Singer, 1993; Regan, 2004), the traditional Yoruba believe only in the welfare of animals and not in the rights of animals. This is understandable because of the various functions they derive from the care of animals (Olusola, 2006, p. 170). Their value and respect for animals does not imply any animal rights or equality of moral consideration. There are many tales and poems by the moonlight that illustrate how the Yorùbá value animals; they are aware of the vital roles that animals play in their lives. By means of folktales and moonlight poetry and stories, they try to inculcate caring for animals in their children from childhood on. For instance:

Adìe mi	My rooster
Èyí tí mo rà	The one that I bought

Motoju re osi dagba	I cared for it till it grows
Losan ojan kan, ó sì je lo	It went out
Lálé ojó kan	One night
Ó kó sí kòtò	It fell into a pit
Ìyá bá mi gbé e	Mother helped me carried it
Gbígbé tí mo gbé e	As I carried it
Gbígbòn ní n gbòn	It was shaking
Mo wá fi yéná	I put it by the fire side for warmth (Olusola, 2006, p. 167).

This children's poem, like many others, is a non-religious moral education. Poetry is commonly used by the traditional Yoruba to educate children on general aspects of life. In the context of the poem, it is expected that when children become adults, they will know that animals should be cared for. It is wrong to make a facile generalisation that the Yorùbá do not care for animals, simply because of the instrumental values they derive from animals such as using them as food sources, for medicinal purpose, and as sacrifices to the deities. A proverb that points to this fact is:

iii *Bí èrèké omo eranko kò bá bàjé, ti omo ènìyàn kò ní dùn*

(If the cheek of the offspring of an animal is not broken, that of the humans' offspring will not be sweet).

This adage points to the idea that, though the killing of animals for consumption is painful, it is necessary for human enjoyment and pleasure. So for the Yorùbá, the sentient nature of animals implies that their welfare should be cared for, especially for the sake of promoting other (not only human but also that of the invisibles) interests and well-being. The tendency to interpet such a perspective as thoroughly utilitarian in outlook is high.

It might perhaps be argued that the Yorùbá are contextualist and have a dose of anthropocentric consequentialism in their relationship with the animate constituents of the environment. They are contextualists in the sense that they do not have a uniform value and attitude towards both animate constituents, flora and fauna parts of nature and the vegetative/plant composition of the environment. Value and care is attached based on the profession the human agent belongs to. A hunter, for instance, will value and care for game in a different way from an herbalist and vice versa.

For the Yorùbá, there is no absolute acceptance of the intrinsic worth of every component of the environment and animals in particular, as Taylor has advocated. As a consequence, responsibility to animals is defined according to one's professional constituency, secular, religious and other considerations. The Yorùbá are anthropocentric consequentialist because their care for animals is based on their understanding of the consequences of irresponsible neglect on human well-being. With such conviction, and on prudential grounds, they attach values and care as humanly possible as they can, for other members of the eco-community in the hope that such an environmental attitude will consequentially promote human interests in the long run.

Within the Yorùbá conceptual system, ideas such as: 'non-human animals,' 'speciesism,' 'veganism,' 'animal rights,' and 'animal companionship' are eccentric. However, this does not suggest a lack of moral consciousness towards non-human animals. Like the animal liberation/rights theorist in the West who extends moral duties to animals on the basis of sentience, the traditional Yorùbá also recognised the sentient nature of animals with some ontologically rooted respects. However, unlike the animal liberation/rights theorist who contends that the only right way to treat animals is to treat them as ends in themselves, and not as a mere means, the Yorùbá recognise only the welfare of animals and not the rights of animals. Morality among the Yoruba is oriented towards communality, duties and obligations, as opposed to rights and individuality. Because of this, "establishing that animals have rights might, consequently, do less in practice" (Ahteensuu, 2016, p. 292) in Yoruba culture. Their religious system reveals that the sentient nature of animals implies that their welfare should be cared for, especially for the sake of promoting others' (humans and the deities) interests and well-being. Among the traditional Yoruba, as in many other traditional African societies, "animals are hardly ever seen as valuable in and for themselves" (Horsthemke, 2015, p. 13).

A promising way of formulating an African animal ethic would be one that posits that the Yorùbá-African worldview advocates an ontological animal care ethics. Ontologised ethics defends the notion that morality is grounded on nature and being. Ontology deals with what entities exist or can be said to exist and how such entities can be grouped, related within a hierarchy. We use the term ontology to characterise the strictest sense of metaphysics that we speak of as the concept of being (Unah, 2005). An African ontology of force injects religious elements into the consideration of principles of right actions and good character in African normative reflections; such religious elements do not necessarily lead to a non-secularist orientation. Having binary complementarity as its conception of being, ontologised ethics seeks to create an equilibrium and harmony among beings within their community by inculcating in the minds of Africans the moral sentiment of togetherness and mutual interdependence among humans, animate and inanimate things. In this rational framing, a moral agent ought to promote a peaceful co-existence, harmony and equilibrium not only between persons with the other beings in the community, as Metz (2007) argued, but also on the whole, between the visible non-hierarchical components of the environment (plants, mountains, rivers, human beings, non-human animals) and the invisible part of the environment (the dead, the unborn, the spirits, etc.).

Though anthropocentric in some sense, the caring aspect of ontological animal ethics is discernible in the Yoruba worldview in the ways by which animals are given meaningful names similar to the practice of naming human beings, and they may be the subject of panegyrics or songs of praise before even using them as a means to an end. To this end, religious beliefs, taboos, folktales, proverbs and moonlight poetry are essentially instrumental in serving as restraints to the mistreatment of animals in Yoruba culture.

These sources equally serve as the platform for oral education on animal care and our obligations towards them. While it is arguable that the challenges of non-human animals that confronted the traditional Yoruba (nay, African) society were not of the same degree as what obtains currently, the fact remains that the reality of the current global experiences suggests that relying purely on traditional environmental values and moral principles may not be sufficient for our contemporary world. The aspects of environmental education and collective-actions, which are embedded in traditional Yoruba environmental practices, can be illuminating for contemporary world. While oral tradition, together with its associated ecological knowledge, is being lost at an increasing rate (because of the expansion of markets, global communications and other aspects of globalisation that promote dominant languages at the expense of native ones), the role of language and culture in the formulation of animal ethics, law and collective activism cannot be undermined.

Conclusion

Thus far, the contentious question of what ought to be the moral status of non-human animals has been engaged in this chapter with a focus on the traditional Yoruba people of Nigeria, West Africa. Though the traditional Yoruba have been regarded as a highly religious people, it is relevant to point out that their religiosity is not an attenuation of the secular and the moral basis of animal care dominant among them. While theoretical ethical reflections on and discussions about the attitudes and cultural dispositions of the Yoruba towards animals have rarely been attempted before now, this chapter attempted such an exploration. Through some hermeneutic consideration of aspects of Yoruba oral tradition, the chapter argues that many of the predominant concepts in environmental discussions are eccentric among the Yoruba. Concepts ascribing and embedding non-human animal moral status such as 'animal rights,' 'animal companionship,' 'equal animal moral consideration' and 'speciesism,' among others, are incongruent to the understanding of non-human animals among the traditional Yoruba. While animals do not have moral claims, there are moral obligations among the Yoruba to protect the animals from undue harm and care for the welfare of animals on the strength of ontological animal care ethics. Such an ethics is underpinned by the principles of good deeds (doing that which is benevolent), not causing pain, respect for consecrated capacities to live, and sustenance of relational attitudes (Owoseni and Olatoye, 2014).

However, the influx of foreign ideas about the treatment and advocacy for non-human animals are increasingly whittling down the predominant traditional perception and attitudes towards them, for ill and for good. Without necessarily despising the moral values in traditional Yoruba care for animals, and alienating the contemporaneous confluence of ideas on how best we ought to protect non-human animals, this work sees as attractive a synergy between old ways and new experiences in order to cross-pollinate ideas for best practices. It is thus important that future studies engage in a reconstruction of the relevant ideas in traditional

Yorùbá ontological animal care ethics for possible integration with contemporary best practices that recognise standards, guidelines and legislations on animal care.

Notes

1 A recently published work that navigated the fundamental question of the moral worth of "other-than-human animals" in African ethics is Kai Horsthemke's *Animals and Animal Ethics*. (Hampshire: Palgrave Macmillan, 2015).

2 Yorùbá is one of the major ethnic groups in Nigeria, West Africa. The Yorùbá people occupy South-West Nigeria and are widely spread across some other parts of Nigeria. They are also found in large numbers in different parts of Africa and in the diaspora: the Republic of Benin, Togo and Dahomey, South Africa, West India, South America and the Caribbean, such as Brazil and Cuba, among others.

References

Adewoye, S. (2007). *Legal Framework for Animals and Game Management in Nigeria*. Ibadan: Positive Press.

Ahteensuu, M. (2016). Kai Horsthemke's Animals and Animal Ethics. *Theoria*, 82(3), 288–294.

Ajibade, G. O. (2006). Animals in the Traditional Worldview of the Yoruba. *Folklore*, 10(30), 155–172.

Attfield, R. (2009). Biocentrism. In J. B. Callicott and R. Frodeman (Eds.), *Encyclopedia of Environmental Ethics and Philosophy* (pp. 97–100). Detroit: Gale Cengage Learning,

Awolalu, J. O., and Dopamu, P. A. (1979). *West African Traditional Religion*. Ibadan: Onibonoje Publishers.

Callicott, J. B., Parker, J., Batson, J., Bell, N., Brown, K., Moss, S., Poole, A., and Wooding, J. (2011). The Other in *a Sand County Almanac*: Aldo Leopold's Animals and His Wild-Animal Ethic. *Environmental Ethics*, 33, 115–146.

Camp, J. C. V., Olen, J., and Barry, V. (2015). *Applying Ethics: A Text With Readings*. Australia: Cengage Learning.

Engel, M. (2008). Ethical Extensionism. In *Encyclopedia of Environmental Ethics and Philosophy* (Vol. 1). Detroit, MI: Gale Cengage Learning.

Fayemi, A. K. (2010). Proverbism and the Problem of Rationality in African Philosophy. *Uche: Journal of the Department of Philosophy*, 16, 64–91.

Fayemi, A. K. (2012). Philosophical Problem of Evil: Response to E. O. Oduwole. *Φιλοσοφία: International Journal of Philosophy*, 41(1), 1–15.

Goodpaster, K. (1978). On Being Morally Considerable. *Journal of Philosophy*, 75, 308–325.

Horsthemke, K. (2015). *Animals and Animal Ethics*. Hampshire: Palgrave Macmillan.

Mangena, F. (2013). Discerning Moral Status in the African Environment. *Phronimon*, 14(2), 25–44.

Metz, T. (2007). Toward an African Moral Theory. *The Journal of Political Philosophy*, 15(3), 321–341.

Murove, M. (2004). An African Commitmment to Ecological Conservation: The Shona Concepts of Ukama and Ubuntu. *Mankind Quaterly*, XLV, 195–215.

Ogunade, R. (2005). *Environmental Issues in Yoruba Religion: Implications for Leadership and Society in Nigeria*. www.unilorin.edu.ng/publications/raymond/envionmental %20issues%20in%20yoruba%20religion.htm (accessed July 12, 2013).

Olusola, A. G. (2006). Animals in the Traditional Worldview of the Yoruba. *Folklore*, 10(30), 155–172.

Oruka, H. O. (1997). *Practical Philosophy: In Search of an Ethical Minimum*. Nairobi: East African Educational Publishers.

Owoseni, A. O., and Olatoye, I. O. (2014). Yoruba Ethico-Cultural Perspectives and Understanding of Animal Ethics. *Journal of Critical Animal Studies*, 12(3), 97–118.

Regan, T. (2004). *The Case for Animal Rights*. Berkeley, CA: University of California Press.

Singer, P. (1989). Animal Liberation. In T. Regan and P. Singer (Eds.), *Animal Rights and Human Obligations* (pp. 148–162). Upper Saddle River, NJ: Prentice-Hall.

Singer, P. (1993). *Practical Ethics*. Cambridge: Cambridge University Press.

Taringa, N. (2006). How Environmental Is African Traditional Religion? *Exchange*, 35(2), 191–214.

Taylor, P. (1992). *The Animal Rights/Environmental Ethics Debate: The Environmental Perspective*. E. C. Hargrove (Ed.). New York: State University of New York Press.

Unah, J. (2005). *On Being: A Discourse on the Ontology of Man*. Lagos: Fadex Publishing.

7 Women and the environment in Africa

Workineh Kelbessa

Introduction

Although several writers have written on the role of indigenous environmental knowledge in general (for instance, Omari, 1990; Kalu, 2001), the contribution of African women to environmental protection has not yet been well articulated. The role of women in development (WID) was the concern of governments and international development organisations in the 1970s. This was replaced by debates on women, environment and development (WED) in the 1980s. Academics and development professionals and activists emphasised the gender and development approach. Scholars have thus gradually focused on gender, environment and development (GED).

> This opened up more constructive opportunities for a better understanding of and engagement with women, gender relations, and environment. This change heralded a more radical approach, the rationale being that through empowerment, women can achieve equity and equality in society. To achieve this, both men and women must be involved, and interventions should be based on a clear understanding of relations, roles, responsibilities, participation in decision-making processes, and so on.
>
> (Flintan and Tedla, 2010, p. 2)

Some ecofeminists[1] also discussed the connections between the domination of nature and women by men. Moreover, they believe that the presence of "feminine" values such as reciprocity, nurturing and cooperation both among women and in nature shows the connections between women and nature. A detailed analysis of this debate is beyond the scope of this chapter.

Like men, women are the custodians of indigenous environmental knowledge. They are also, among other things, guardians of the home. In this chapter, I explore the contribution of African women to environmental protection by taking some examples from the African continent. Although there are visible practical differences in different countries, African women share common practices and beliefs. In spite of the fact that the cases examined are largely from Oromia, Ethiopia and Kenya, most of the issues are relevant for other regions of Africa.

The Oromo constitute one of the largest ethnic groups in Ethiopia. They belong to a Cushitic group in East Africa. Other Cushitic speaking groups include Somali, Konso, Afar, Sidama, Kambata, Hadiya, Darassa, Agaw, Saho, Beja, Gedeo, Burji and others. The Oromo are also found in Kenya. Currently, Oromia is one of the regional states of the Federal Republic of Ethiopia.

The Oromo people have developed the *gadaa* system, an indigenous democratic socio-political system (for details, see Legesse, 2000; Kelbessa, 2011). The *gadaa* system has its leaders who serve their people for an eight-year term. It organises the Oromo society into groups or sets (about 7–12) that assume different responsibilities in the society every eight years. The *Gumii/Ch'affe*, the national assembly, formulates different laws, including family, political and animal laws. These laws are subject to change in response to new developments and challenges. It should be noted that the United Nations Education, Science and Cultural Organisation (UNESCO) inscribed the *gadaa* system as an intangible world heritage in 2016.

As I will show later, women are excluded from political leadership, although they are consulted for decision-making regarding the protection of the rights of women. Oromo women also have their own organisation that helps to promote group solidarity and protect their rights. It is hoped that the accumulated experiences of Oromo women can be beneficial to other women who fight for their rights in Africa.

Part one looks into the link between women and the environment. Part two examines how African women strive to protect their interests and rights. The last part provides concluding remarks. This chapter does not set out to provide a comprehensive account of the impact of African women on environmental protection. Its aim is to stimulate rather than to instruct, to open up rather than to close off options, and to make a modest contribution to the subject.

Women and the environment in Africa

Although women held positions as chiefs, had their own organisations and participated in male councils in some cultural groups in pre-colonial Africa, women in other cultural groups were subject to different aspects of subordination because of customary laws. For instance, there were women chiefs in the Tonga people (Coquery-Vidrovitch, 1997, p. 34), in West Africa, in pre-German central Cameroon, among the Flup of Casamance in southern Senegal (Jean-Claude Barbier, 1985, cited in Coquery-Vidrovitch, 1997, p. 35), and "among the Mende and Sherbro of Sierra Leone" (Coquery-Vidrovitch, 1997, p. 35). Between 1914 and 1970, ten of the 146 recognised chiefs in Sierra Leone were women (Hoffer, 1972, cited in Coquery-Vidrovitch, 1997, p. 35). There were also female regents in Africa (Coquery-Vidrovitch, 1997, pp. 37–39). A case in point is Empress Menetewab (ca. 1720–1770) of Ethiopia, who became regent in her son's name following the death of her husband in 1730 (Richard Pankhurst, cited in Coquery-Vidrovitch, 1997, p. 37). Moreover, Queen Dola, a half-sister of Akengbuwa, *olu* (chief) of the Itsekiri Kingdom (in present day Nigeria) established a state council to maintain the "continuity of the regime," although she died in 1848 without fulfilling her aim (ObaroIkime, 1970, cited in Coquery-Vidrovitch, 1997, p. 37). During the pre-colonial period, African

> [w]omen varied from being highly subordinated 'legal minors' under the control of their menfolk among groups like the Tswana and Shona in southern Africa to holding positions as chiefs among the Mende and Serbro of Sierra Leone and 'headmen' among the Tonga of Zambia. In some societies, women even had formal roles in male councils. The figure of the queen mother in many societies of western Africa was very influential, and it was she who selected the king.
>
> (Gordon, 2013, p. 307)

Moreover, both men and women had their own respective parallel authority structures that enabled them to undertake their own actions and "exercise authority over their own sex" (Staudt, 1987 p. 195; Parpart, 1988 pp. 208–210, both cited in Gordon, 2013, p. 307).[2] However, this does not mean that men and women had equal power. The authority of men was more formal than that of women. "For most women, power was (and still is) exercised indirectly and informally as sisters, mothers, and wives within the extended family system and was closely associated with women's economic power" (Gordon, 2013, p. 307).

The colonisation of Africa promoted the commercialisation of agriculture and thereby changed the customary gender division of labour (Gordon, 2013, p. 309). Men specialised in cash crops for export, whereas women continued to produce food crops. Colonial authorities required men to cultivate new crops (peanuts, cotton, cocoa, coffee, etc.) during the early colonial period (Coquery-Vidrovitch, 1997, p. 59). Men moved to towns and worked in different fields. For example, some Swazi men were employed in the mines and settler farms in South Africa, whereas Swazi women produced food crops for their families (Kuper, 1947, cited in Sikhondze, 2003, p. 84). Most women continued to work in rural areas and had no access to money, skills and education available in urban centres. Far worse, the new colonial rulers promoted the authority of men over women. They created all-male "native authorities" that facilitated men's control over women (Chanock, 1982, cited in Gordon, 2013, p. 309).

Post-colonial African governments have adopted similar policies and African women have continued to face similar problems, despite some notable exceptions in some countries.[3] Some men have continued to migrate to neighbouring countries. For instance, most women in south-west Zimbabwe are "*de facto* family heads for most of the year," as their husbands work in Botswana or South Africa (Matsa and Matsa, 2010, cited in Mark and Winniefridah, 2013, p. 325). Divorced and widowed women become *de jure* family heads.

Even after independence, some African men have continued to aggravate biodiversity loss by promoting cash crops at the expense of food crops and undermining women's empirically based knowledge, roles, perspectives and needs. According to Jacobson (1993), "[i]n much of sub-Saharan Africa, for instance, both men and women plant crops, but they do so with different goals. Men grow cash crops. . . . Women, by contrast, use their land primarily for subsistence crops to feed their families" (p. 64). Shiva (1995) also writes, "[i]n Africa, basic food production continues to be in the hands of women, even as their control over agriculture is increasingly being eroded through [the] green revolution and cash-crop farming. Women there do 70–80 per cent of all agricultural work and produce 40 to 50 percent of all staple food crops" (p. 108).

According to Seager (2009), "up to 80% of basic foodstuffs both for household consumption and for sale" are produced by sub-Saharan Africa women (p. 69). Another study reports that 40% or more of the agricultural work force of 46 of Africa's 53 countries are women (Nabalamba, Mubila and Alexander, 2011, p. 1). Amadiume (1987) also stated that in the Igbo of Nigeria women are expected to produce crops with less or no commercial value, whereas men focus on the production of cash crops. The commercialisation of agriculture has increased women's workload for producing subsistence in Africa.

In spite of the hardships women have faced, they have played an important role in the promotion of biodiversity conservation in sub-Saharan Africa. They have managed energy, water, farming, livestock, forests and the like. They have selected plant genetic resources on the bases of different criteria including nutrition, taste, texture, processing requirement, storage quality, resistance to pests, adaptability to soil and climate and so on. According to Jacobson, women "have played a leading role in maintaining crop diversity. In sub-Saharan Africa, for instance, women cultivate as many as 120 different plants in the spaces alongside men's cash crops. . . . Women in subsistence economies also are active managers of forest resources, and traditionally play the leading role in their conservation. Forests provide a multitude of products to households. They are, for example, a major source of fuel" (1993, p. 68). Women also plant trees and collect wild foods to supplement their families' diets. Some rural women know the medicinal value of some trees, as they gather barks, roots and herbs for medical reasons.[4]

Like their male counterparts, Oromo women in Ethiopia have also contributed to environmental management and biodiversity conservation.[5] They are concerned about the well-being of humans, non-human beings and the natural environment. They pray to *Waaqa* (God) around a sacred *Qiltu* (sycamore tree), springs, on a mountain or around other sacred trees during ecological crisis. As I have stated elsewhere (2011), the Illuababorra Oromo, in most cases women, used to sit under *qoloo* trees, which are believed to be indwelt by spirits and

> pray to *Waaqa* for rain, for protection from sickness and death, famine, hail, drought, and failure of crops, for gifts of longevity, children, prosperity, peace and for protection from evil spirits and wild animals. The Oromo are not worshipping the material symbols but the spirits of the symbols represented by trees. They proudly say that their Creator does not want to see the death or destruction of His creatures, and that is why He has come back with the rain of life.
>
> (p. 83)

Hussein (2004) recorded one rainmaking *ateetee* hymn of the Arsi Oromo women in central Ethiopia (pp. 111–112). It is believed that, after the women have prayed, God will give rain to the community. According to Hussein (2004), "[t]he significance of this mythology is that it demonstrates the leadership role of women in defense of the community" (p. 112).

The Borana Oromo pray for the fertility of cattle and of women, multiplicity of offspring and of cattle, for peace, rain, pasture and health.

> The peace of the Boran consists not only of preventing internal strife and fostering active co-operation between men, but also maintaining proper relationship between men and God, so that the blessings of the latter, which are manifested particularly in rain and in the fertility of stock and women continue to flow.
>
> (Baxter, 1978, quoted in Shongolo, 2004, p. 51)

In this connection, it is worth noting that the Oromo and other African women have played an important role in maintaining peace. The Oromo women participate in the *gadaa* rituals and peace negotiations as a link between the opponents. For instance, when there is a disagreement between the Borana and Guji Oromo, in south Oromia, "[t]he party first interested in peace sends a *lichoo*, a female peace envoy, to the *hayyu*, judge, of the opponent group. While traveling in the territory of the enemy, the custom confers respect to a *lichoo* and protects her from any harm. Even if the adversary is on a war campaign, the group returns back at the sight of a *lichoo*" (Debsu, 2009, p. 25). For minor conflicts, a man can also serve as a *lichoo*. It would be wrong to attack a *lichoo*, who holds a whip, in the territories of an enemy, as her or his purpose is to maintain peace between the conflicting parties. Similarly, Somali women can serve as a bridge between the conflicting clans in Somalia. "If one party refused to participate in a peace agreement, the elders often used a woman who belonged to one group but had closer relations in the opposing group to open contact" (Olsson, 1994, p. 73).

Oromo women also sing songs in praise of their cattle. Consider the following songs:

Akka kinnisaa dammessitani
Boosonillee hin woyyessitani
Ee saayya hoofaayya meeshaa mormaa
Loonii toltu lafa daarii qonnaa

You make honey like a bee
You improved life for even slothful ones
Oh Saayya, beautiful like a necklace
Only when it is good for the cattle
Are we able to do farming to the fullest extent (Wako, 2011, p. 45).

Yaa loowwanii koomanee
Ulfinni eessa jira loon malee?

Oh! Cattle, we are helpless
Where is the glory without them? (Wako, 2011, p. 46).

Moreover, the following *suunsuma* (war song) illustrates the place of cattle in Arsi and the close relationship between women and cattle:

Heefa as naa waamaa,	Call the women here.
Wonna loonii gaxxamaa,	I will recite a poem about cattle
Wonna looni fi heefaa,	About women and cattle
Jarri wolil gosho'o,	The two are close friends
Loowwan irraa oofnaan,	If you take away the cattle from them
Ati deebiin goto'o,	You will never be allowed back here again
Qofa loon hintuqinaa,	Do not ever harm the cattle
Achi duubaan coko'o.	If you do, the price is 'coko'
Jara maqaa hin dhoofnu,	We do not name their name
Arraba dooniitii,	A taboo of the tongue
Wonti Arsiindadhabu bee	What the Arsis cannot tolerate is harm done to the cattle
Wonnuma looniiitii,	(Tegegn, 1994, quoted in Wako, 2011, p. 1).

Coko is one of the Arsi spears. The number of heads of cattle one owns determines one's richness and greatness in Oromo society.

In the Oromo worldview, the killing of women and children during war is strictly forbidden. "Far from being an honourable act, to kill a woman would be spoken of with horror" (Bartels, 1983, p. 284). Furthermore, it is strictly forbidden for Matcha and other Oromo women to kill or witness a killing. The point is that the woman is a source of life and should not destroy life.

Women have also contributed a lot to food production and processing. They have used various indigenous food preservation techniques. Milk processing is one example (see Mark and Winniefridah, 2013, p. 326).

Women have also played an important role in livestock development and management. They raise animals and know the value of different breeds. Mark and Winniefridah (2013) reported that in south-west Zimbabwe, "[a]ll respondents conceded that care of young ones and the sick lambs and calves is the responsibility of females [rather] than males. The orphans and the sick can be given porridge or assisted to feed (suckle) from other cows until they are able to feed on their own. This can be done by both males and females. In case of rains and colds, women keep these in kitchens with fire for warmth" (p. 332).

What should be stressed is that women are expected to undertake household and reproductive tasks, including growing crops, cleaning, cooking, shopping, childbearing and care giving, the provision of water, fuel and other products, manufacturing household items for their families, collection of fodder for livestock, the management of young calves, looking after sick and weak livestock and milking[6] livestock. Borana women in Ethiopia and Maasai women in Kenya (Njiru, 2013) are also required to construct huts. As I stated elsewhere (2011), a clear division of labour by gender exists in Oromo society.

> For instance, in Borena, military and political activities, the construction of kraals and fences, preparations of wooden utensils, the building of dams,

> tilling the land, taking care of herds and excavation of wells are believed to be the vocation of men. Thus, men are more knowledgeable than women regarding these activities. Women cannot participate in the elections of leaders of camps and *Gadaa* classes and defence of wells, herds and shrines. They do not bury corpses. On the other hand, the construction of houses, and the preparation and distribution of food are delegated to women. Women can tend animals. However, men can participate in the construction of houses with corrugated iron. Both women and men can participate in various forms of ritual although men decide on these issues. In Borena, it seems that women have more responsibilities than men do.
>
> (pp. 159–160)

Oromo women also participate in war by digging trenches, raising ramparts, fortifying camps and providing supplies (Phili Caraman, 1985, cited in Etefa, 2012, p. 85). According to Johann L. Krapf, "The weapons of the [Oromos] are a spear, sword, and shield, and they all ride on horseback; even the women gallop beside or behind their husbands" (p. 62, quoted in Etefa, 2012, p. 85).

Lowland Guji Oromo women make coffee, construct houses, make beds, clean calves and goats kraal and all residential areas, collect firewood, mow grass for house covering and fodder, sell animal products and foodstuffs in the market, and carrying maize from the field to storage facility, in addition to their domestic activities (Debsu, 2009, pp. 25–26). In the past, Guji men also milked cows (Debsu, 2009, pp. 25 and 26).

In southern Malawi, men are required to construct houses, cultivate land and burn charcoal, whereas women are expected to draw water, fetch firewood, fish, cook, wash, clean and take care of children. Both men and women can cultivate, cut grass, and plant trees (Mawaya and Kalindekafe, 2010, p. 103).

Both men and women employ different strategies in response to water stress and droughts. Pastoralists use different coping mechanisms, including livelihood diversification, drought-tolerant crops, like sorghum and millet, herd diversification, drought-resistant animal species, crop cultivation, seasonal mobility, relocation of livestock to better pastures, using wild fruits and dried plants as human and animal feed, digging up and chopping tubers for livestock, harvesting fleshy creepers from rocky outcrops, fishing, petty trading[7] and the like to deal with the problems of climate change (see Amsalu and Wana, 2013; Balehegn and Tafere, 2013; Mark and Winniefridah, 2013; Njiru, 2013).

Besides some of the above mentioned strategies, Afar women in Ethiopia have used dried meat (*Darru*) and evergreen wild food such as *Amberaarcu* and *Doberaglabra* (Fassil, Diress, and Synnevag 2001; Nagar 2001; Wawire 2000, cited in Balehegn and Tafere, 2013, p. 90). Pastoralists in different parts of Africa favour browsers (camels and goats) over grazers (cattle and sheep) to cope with water scarcity and the encroachment of bushes and shrubs that have made their rangelands unusable. "Camels and goats give milk even during dry periods, when lactating cows can hardly be milked" (Amsalu and Wana, 2013, p. 35). Maasai women pastoralists also make beads, sell milk, burn and sell charcoal, and sell firewood and land (Njiru, 2013, p. 66).

African women have suffered from longer hours of working, indoor air pollution because of poorly ventilated rooms when they cook with firewood, discrimination in decision-making processes because of traditional gender roles and indigenous patriarchal excesses. Women in sedentary agricultural and pastoral communities are the major victims of climate change and environmental degradation because of their diverse responsibilities. Among others, they are required to fulfil the basic needs of their families. Gordon (2013) states,

> Women are also disadvantaged by 'time poverty' and a lack of investment in women's productivity. Studies consistently show, for instance, that women work much more than men. Not only do women work long hours in the field (8 hours for women vs. 6 hours for men), they spend additional hours processing food; cooking; and collecting firewood, fodder for animals, and water. Additional time is spent taking surplus produce to markets to sell.
> (pp. 312–313)

Similarly, Kipuri and Ridgewell (2008) explained the working condition of pastoralist women as follows:

> Pastoralist women must work longer and harder than men, fulfilling 'female' roles in the household, as well as making money from tasks traditionally deemed to be 'women's work', including collecting firewood, and making and selling handicrafts. This labour is in such demand that girls are often removed from school in order to work. Women do not attend many of the social occasions at which men make decisions that affect the whole community.
> (p. 3)

As stated earlier, some women are required to construct houses. Materials for construction, such as thatching grass and wood, are declining because of environmental degradation, which in turn increases burdens on women, as they are required to travel long distances in search of construction materials, firewood and the like.

Debsu (2009) also reports that the transition from pastoralism to agropastoralism increased women's work burdens and reduced their ownership rights in Guji Oromo in Ethiopia. Men rest during the dry season. However, women have no time to rest, as they "remain busy throughout the year with routine domestic chores and agriculture related activities although they have little control over agricultural products" (Debsu, 2009, p. 26). The only exception is that, in the past, Guji women were not expected to engage in routine domestic and extra-domestic activities "for about six months after giving birth to children" (Debsu, 2009, p. 26).

Furthermore, some studies have shown that women suffer more than men from the effects of climate change (Denton 2002; Intergovernmental Panel on Climate Change [IPCC], 2007; United Nations Framework Convention on Climate Change [UNFCCC] 2007; Aklilu and Alebachew 2009; Anderson, Hesse and Nassef 2009, cited in Mulinge, 2013, p. 6). Severe droughts force men to migrate to other places, whereas women stay behind and take care of children and weak livestock. The workload of women has been increased by climate change.

Moreover, in their study of the status of women in Southern Malawi, Mawaya and Kalindekafe (2010) noted that lack of literacy and education make women "vulnerable to all sorts of hardships, and the fact that women were not involved in decision-making processes placed them in an even more vulnerable position" (p. 123).

The male-dominated culture of various ethnic groups in Africa has established the basic relations of power and undermined the intellectual development of women. Men have used proverbs and wise sayings to characterise women in negative terms. According to Balogun (2005), "proverbs have served as a major avenue for the continued perpetration of gender discrimination among the Africans" (p. 39). It is believed that women are deceivers, senseless, morally debased, irresponsible, devilish, childish, ignorant, wicked, dangerous and irrational. On the contrary, men are portrayed as knowledgeable, rational, strong, heroic, self-reliant, competitive and true leaders of their society. As such, men have employed proverbs to emphasise the difference between feminine and masculine roles. These in turn have forced women to accept a lower position in their societies. Consider the following proverbs:

"**The hen knows when it is morning, but she looks at the mouth of the cock**" (Kiiru, 1999). (Ashanti proverb, Ghana)

"**The woman is never in charge**" (Schipper, 2006, p. 252). (Igbo proverb, Nigeria)

"**With a woman's rule, the gate remains unopened the whole day**." (Oromo Proverb, Ethiopia)

"**If a cow leads the herd, all the cattle will fall into the pool"** (Schipper, 2006, p. 235). (Venda Proverb, South Africa)

"**He who listened to the woman suffered famine at harvest time**" (Kiiru, 1999). (Bemba proverb, Zambia)

"**Obinrinko se finuhan**" "Secrets should not be revealed to a woman" (Ajibola, 2005, p. 26). (Yoruba proverb, Nigeria) This proverb implies that it is woman's nature to divulge secrets.

"**Woman and sky cannot be understood**" (Schipper, 2006, p. 295).[8] (Gikuyu proverb, Kenya). This proverb conveys that both women and sky seem to be mysteries that often change.

As stated earlier, women in many African countries are excluded from decision-making processes. Among others, in the Oromo of Ethiopia, women are not included in decision-making processes. Very often development and environmental decisions are made without taking into account the voice and concerns of women.[9] Women cannot be the *Gadaa* leaders of their society. "They do not participate in the *Gadaa* system partly because of their dual identity – their own and their husband's. As lineage is counted through a male line, women's identity does not count. For this reason they cannot go through all the *Gadaa* grades" (Kelbessa, 2011, p. 160). By rule, it is impossible for *gadaa* officials to leave the *gadaa* centre and live somewhere else. Women can leave their place and live in other places. This is one of the reasons why women cannot join the *gadaa* leadership.

Some elders argue that *Waaqa* (God) qualifies the female sex for the bearing and rearing of children and maintaining the family and household (for instance, cooking, collecting water and fuel). The following Oromo proverb indicates the inherent inferiority of women. "**Nadheen furdoo qabdi malee beektuu hinqabdu**" – "**One can find many fat women but not knowledgeable ones.**" However, this is not a defensible view. Cultural domination has forced women to be subservient to men.

Furthermore, although pastoralist women have contributed to the environment and drought mitigation strategies, policy makers do not include the views of women in climate mitigation measures. According to M. Mwangi "drought policies are designed only by men: these policies are silent about the vulnerabilities of women to climatic hazards. Ignoring the realities and needs of half of their societies undermines . . . the viability and effectiveness of any drought response" (p. 22; quoted in Kipuri and Ridgewell, 2008, p. 18).

Women's struggle in Africa

In this section, I will discuss how African women have attempted to protect their rights in Africa. They have shown the importance of African women's knowledge and indigenous institutions for the protection of women's rights and interests. Some of them have become role models for oppressed women. Their courage and achievements should serve as inspiration to all who fight for justice.

African women have used different strategies to protect their interests. One of these strategies is using stories to challenge and deconstruct masculine and imperialist points of view. According to Louise Du Toit (2011),

> Resisting western philosophy's tendency to subsume African women under the notion of 'Man', African women tell stories about their situatedness in the world. The narrative within this tradition do not only simply or merely 'reflect' an African women's world, but is to a large extent the founding political gesture with which African women deconstruct masculine and imperialist points of view on the world which claim to be neutral, but which in reality conform largely to masculine and colonial or neo-colonial desires and needs. (p. 123)

The Oromo women have relied on indigenous institutions to promote group solidarity against the abuses of husbands and other individuals. According to some writers, Oromo women held a relatively better position in the past because of the *ateetee* ritual (Hussein, 2004) and the *siiqqee* institution (Kumsa, 1997). It should be noted that the Arsi Oromo use the name *siinqqee*. Others use the name *siiqqee*. I use these words interchangeably. "*Siiqqee* is a stick symbolising a socially sanctioned set of rights exercised by Oromo women" (Kumsa, 1997, p. 115). The *Gadaa* system incorporates the *siiqqee* institution that was designed to protect the rights of women. Oromo women use *siiqqee* to protest against the violations of secular and religious rights, and infringement of their rights by different individuals and groups, and to restore order and harmony.

The Oromo regard women as both outsiders and strangers in the *gosaa* (clan) into which they were born and into which they were married respectively. They are described as the weaker sex (*nadheen imuka laafa*: women are "soft wood – a depiction of their liminality)" (Kumsa, 1997, p. 119).[10] Women are analogised to delicate sticks that need special care and treatment in all social relations.

The *siiqqee* institution has the role of protecting women, the "softer" and the weaker half of the population, and maintaining balanced relationships between women and men. Similarly, there exists a parallel organisation called "*gaaseyba*" in the Orma Oromo of Kenya (Kelly, 1992, cited in Kumsa, 1997, p. 119).

The word "*ateetee*" has different meanings. On the one hand, it refers to a spirit that is believed to be useful for animals and human beings. For the Ambo Oromo of western Ethiopia, *ateetee* (female divinity by women) is the mother of cattle and the spirit of *baksaa* (melted or processed butter). The Oromo also identified *ateetee* as *Aayyoo Baar* (the Mother of Ocean) and as *Hadha Dambal* (the mother of overflow, full). The purpose of the *ateetee* ritual is to help cattle breed well and to help oxen plough well. The Ambo Oromo perform *ateetee* ceremonies every two years. The *ateetee* ritual has symbolic meanings. The anointment of sticks with butter, the planting of green poles, the shedding of old cows' blood, the splashing of the chest and the neck with butter are the symbols of fertility, procreation and continuation of life on Earth. They symbolise that the survival of most Oromo depend on the survival and rebirth of herds (for details, see Kelbessa, 2001).

The spirit that resembles *ateetee* in Arsi is known as *fatimme*. *Fatimme* is a spirit and is believed to make animals fertile. *Fatimme* possesses a woman on the celebration day. Women also appeal to *fatimme* during labour for help. Women in Arsi stopped celebrating *fatimme*, because *sharia* outlaws it. An extended discussion of this subject is beyond the scope of this chapter.

Ateetee has two different forms in Arsi. The first form of *ateetee* is a peaceful mobilisation of women to pray to God. They hold *siinqqee* sticks, ululate and pray to God under a tree or around a river bed (*malkaa*), or on a mountain, and around springs, particularly during a long dry season, crop failure, endemic diseases, war and other crises. This is a peaceful *siinqqee*.

Secondly, the *ateetee* ritual has had a political function. In Arsi, women used to challenge men who harassed or attacked them. The *ateetee* ritual promotes group solidarity among the Oromo women. Accordingly, the second form of *ateetee* ritual is the mobilisation of women to protect the rights of women when women face various problems, specifically, when a breastfeeding woman with *qanafa* (a hairband) on her forehead is beaten and harassed by her husband, when a son beats his mother, when the family members unnecessarily harass a woman, when the father harasses his son's wife, and when a woman is raped by a man, the victim ululates and asks other women for help. This is called an *ateetee* strike, and the *siinqqee* scream as all women who join their colleague ululate and try to punish the wrongdoer. *Siinqqee* sticks are the spears of women, as they do not have real spears that belong to men. This has nothing to do with *ateetee* (a spirit), known in other parts of Oromia.

When the husband beats his wife who is wearing *qaanafa* or verbally embitters her, she will notify the *sadedeettan hanfalaa* (council of senior mothers) by leaving her house and screaming in the field in Arsi. *Saddeetoo* or *saddeettan hanfalaa* is "a non-kin Arsi Oromo women's social organisation" (Harris, 1995, cited in Hussein, 2004, p. 107). It is worth reiterating that one *gadaa* period has eight years. Likewise, in Arsi, women have eight senior mothers who act on behalf of women and protect them from abuse. The leader of such women is called *qarichoo*. Only the leader will sit down with men and discuss the issue. *Saddeettan hanfalaa* is a kind of solidarity that was developed to allow women to participate in village council and thereby defend the women (for details, see Hussein, 2004). This village sphere association can include only married and postmenopausal women.

Hanfalaa is a leather strap used to tie the waist of women. Women use *hanfalaa* during holidays and other festivals. Women tie their *hanfalaa* and pray to God during crisis, drought and the spread of new diseases. In Borana, women cannot walk without a *hanfalaa*. If the woman travels without a *hanfalaa*, it means that she is walking naked. This is stated in Borana law.

After receiving the message of the mistreatment of their colleague, the council of mothers and all women in the village would abandon their individual house and react against the wrongdoer. They would meet around the premises of the misbehaving husband and sing songs that express discontent (Hussein, 2004, p. 113). Because of the negative consequences of this group solidarity among women, their husbands do their best to respect the rules. If their case is not properly handled, they won't go home and perform their activities at home until the fallen *qanafaa* returns to the forehead. This means that the members of their family will starve until their case is settled. They spend nights outside the house of the wrongdoer. They also go to the house of a woman who failed to join them and begin to ululate. They take everything that can be eaten and drunk from her house and consume it. They also cook and eat what they found in the house. Her husband cannot prevent them from doing so. If he tries to do so, they will force him to sacrifice a cow for them.

Unless the local elders approach the women to make peace, the women will take *bidhaa* (shelter) in the neighbouring clan, which is required to give them accommodation and food. *Bidhaa* is a sanction against "the council of male elders." According to the rule, the council of elders in the boycotted clan should compensate the women by performing reconciliation rituals for failing to handle the case on time.

> The conflict resolution starts with approaching the women and asking for forgiveness for invading their territory. This is called *dhiltee dhinnaa* (seeking

> apology for direful eye contact). As a prerequisite, the male elders, therefore, take some lush grass to recite the *dhiltee dhinnaa* and to show the veneration for the gathering. It is only when this rule is observed that the conflict resolution process begins. As they hand over the grass to the senior mothers, the male elders utter in unison, *dhiltee dhinnaa* (save us from your eyes). As they receive the grass, the women on their part recite in unison, *Haffola Hobbaya* (be safe!) *Ijarraa hafaa*! (survive our eyes!)
>
> (Hussein, 2004, p. 114)

Consequently, the elders ask the women the cause of their strike. After one of the senior mothers describes the issue, the elders would ask the miscreant to sacrifice one of his cows or calves for the gathering. If the offender is willing to do so, the women will forgive him. If the offender fails to admit his mistake and make peace, the women impose a more serious curse called *abaarsa siinqqee* (the curse of *siinqqee*). After putting down their *siinqqee* (ritual sticks) on the ground, the women who gathered there pray to God to eliminate the wrongdoer (Hussein, 2004, p. 114). If the elders cannot solve their problem, the case will be referred to the *gadaa* officials. Then the *gadaa* officials will take the case seriously and solve it. It was very rare that men refused to make peace with women who are supporting the beaten woman. As Baxter notes, "[a] man is not only made uneasy before a group of outraged vociferous women; he has no recourse from them and must either submit or flee" (Baxter, 1979, p. 13, quoted in Etefa, 2012, p. 131).

Furthermore, pregnant women command respect in Oromo society. The *Gadaa* assembly can remove high-ranking *gadaa* officials who mistreated a woman. A case in point is *Abbaa Gadaa* Wale Wacch'u of Borana (1722–1730). Legesse (2000) reports:

> But the final incident that provoked his dismissal was that a pregnant woman had come to his council and asked for water. He and his councillors told her that pregnant cows drink water on all fours and that she should do likewise – hands and knees on the ground. Denying a Borana, who committed no crime, access to water is a crime. Denying a pregnant woman such help is an even bigger crime, because she carries a gift of God. The Gumi of Wale Wacch'u (1722–30) decided to have these *hayyus*, including the *Abba Gada*, chased out of the seat of *luba* government. They were exiled. They said to Wale Wacch'u Warri ch'ufsittiorme! 'All the people have banished you!'
>
> (p. 204)

Both internal and external factors undermined the *Gaadaa* system after the conquest of the Oromo land by Abyssinians towards the end of the 19th century. The role of *Ateetee/sii(n)qqee institution* was reduced to ritual. The introduction of foreign values and religions undermined the values of indigenous institutions.

Like the Oromo women, the Barabaig women in Tanzania have some rights to protect themselves from men. In particular, they use the fertility ritual called *ghadweta* to protect their welfare (Blystad, 1995, p. 95).

After studying the worldview of the Yoruba people in Nigeria, Sofola (1998) maintains that, according to the African worldview, both genders have the same divine source, although they have their own special roles to play in the life of the community. According to Yoruba mythology, "disaster struck when men tried to marginalise the female member of the entourage. When *Orunmila*, the Ifa oracular deity, was consulted, it was revealed that the female member had been marginalised. Quickly, the abnormality was rectified and tranquility restored" (Sofola, 1998, p. 53). According to Sofola (1998), the foreign European/Arabian male-centered systems of authority and governance destroyed and completely eliminated the female lines of authority and socio-political power by de-womanising African womanhood (p. 61).

Aidoo (1998), for her part, indicated that most African societies were matrilineal from the pre-pharaonic period all the way down to a micro-nation like the Akans of Ghana. Aidoo lamented that, first, Islam and, later, Christianity changed matrilineal systems in some African countries, for both were profoundly patriarchal in orientation. Matrilineal societies recognise the sexual rights of both men and women. In Zimbabwe, women who are in this category are still free to have sexual affairs with their lovers in the absence of their husbands' right in their house (Arnfred, 2004, p. 12).

Nzegwu (2006) also states that, before colonisation, Igboland was a non-patriarchal society. Prior to European colonialism and Christian and Islamic influences, women were sexually autonomous and could control their sexuality in Igboland. They were "assertive," economically independent, and industrious. Women in or outside polygynous marriages had the right to have lovers to satisfy their sexual desire.

Although some writers appreciated the place of women in matrilineal societies, one study shows that, even in matrilineal societies, men are the heads of households and the sole decision-makers. For instance, the majority of people in Southern Malawi

> follow a matrilineal descent system and matrilocal settlement patterns in which, after marriage, men live at their wives' homes. One might therefore expect women to be dominant in decision-making at household level. However, this was not the case; despite men coming from the outside, they still played a cultural role of being decision-makers in the house hold. The research data indicated that women respected men and considered them to be household heads.
>
> (Mawaya and Kalindekafe, 2010, pp. 100–101)

In many African societies, women have only usufruct rights of land in both patrilineal and matrilineal societies (Flintan, 2003, cited in Mawaya and Kalindekafe, 2010, p. 113). Although women have the legal right to own land or trees, they do not control them (Mawaya and Kalindekafe, 2010, p. 113).

As Toit (2011) notes, contemporary African women thinkers, including Tanella Boni (from Ivory Coast), Awa Thiam (from Senegal), Sophie Oluwole (from

Nigeria), Marie Pauline Eboh (from Nigeria), and Mercy Amba Oduyoye (from Nigeria) have relied on African values and supported women's struggle against patriarchy and sexism. We can also add Nzegwu, Ifi Anadiume, Ama Ata Aidoo (from Ghana), Rozena Maart (from South Africa), Fatou Sow, Penda Mbow, and Aminata Diaw Cissé (from Senegal), Ariane Djossou (from Benin), Abena Busia (from Ghana), Mary E. Modupe Kolawole (from Nigeria) and the like. Non-African scholars, such as Gail Presbey, Signe Arnfred, Liv Haram and many others have advanced the causes of African women.

It is also worth noting that Wangari Maathai founded the first women-centred environmental organisation called the Green Belt Movement (GBM) in 1977 in Kenya. The GBM links gender and natural resources. It mobilises women to plant indigenous trees and restore degraded landscapes in different parts of Kenya. The GBM has created employment opportunities for women and encouraged them to utilise natural resources sustainably. One hundred thousand women activists participated in the GBM. Six hundred community networks were established. These networks have established 6,000 nurseries. The members of the GBM planted more than 30 million trees in Kenya (Maathai, 2007,175). Planting trees has enabled women to gain better access to clean water and firewood for cooking. Maathai and her colleagues reclaimed public lands given to private individuals by the former President Daniel Arap Moi (Cesar, 2010, cited in Presbey, 2013, p. 282).

Originally, Maathai aimed to save trees and the environment. Later on, she tried to increase women's self-confidence, and help women air their voice by expanding a democratic space (Maathai, 2010).

The GBM established the Pan-African Green Network in 1986 to extend its work to other countries in Africa and the world (TAVANA, n.d.). Thirty African countries and the United States of America hosted the movement by 1997, where over 80,000 people benefitted from its work (Nivola, 2008, cited in TAVANA, n.d.).

Maathai, the Kenyan professor, activist and environment minister, received the Nobel Peace Prize and the French Legion of Honour in 2004 and 2006 respectively. The University of Nairobi also recognised the contribution of Maathai. It established the Wangari Maathai Institute for Peace and Environmental Studies (Presbey, 2013).

Conclusion

In this chapter, I have shown how African women have contributed to the protection of their natural surroundings, food production and processing and the well-being of their families and societies under difficult conditions. Some African women are environmental activists. In many African countries, women do not own and control land and other resources because of gender roles constructed on the bases of customary laws. Male political leaders and institutions have limited the role of women in environmental management. Far worse, many African governments and international financial and donor community have ignored women's interests and rights. Many policy makers in Africa have ignored the knowledge and concerns of women. In spite of this, women have continued to develop and improve their environmental knowledge. In the future, the ongoing deconstruction of the exploitative and hegemonic gender relations within African countries will encourage women to articulate and expand their knowledge, thus contributing to the dynamic development of African traditions.

However, African women have a long way to go to realise their objectives. They have to empower themselves and challenge patriarchal values that perpetuate their exploitation. They should try to make their point in order to change environmentally destructive cultural traditions and participate in environmental decision-making. Responsible African and non-African scholars, both men and women, should support the struggle of women and raise public awareness about the place and role of women in their societies. African governments should also train women so that they can contribute to environmental management. The knowledge of women should be given the attention it deserves. It should be stressed that, compared to men, women are the first to know the decline of the quality and quantity of different local resources because of their responsibilities for the needs of their family members. It is true that, compared to men, the coping and adaptation measures taken by women are weak because of the fact that they do not have equal access to resources that can strengthen their activities. There is no doubt that the activities of women and men are complementary. So, both groups should support each other and promote environmentally and socially sound practices. The integration of informal and formal institutions can help people to improve the quality of their environment and their own well-being.

National, regional and international laws and action plans should be translated into practice, including, for instance, a Gender Policy and Action Plan that aims to promote gender equity and women's empowerment adopted by the African Union Assembly in 2008, the Convention on the Elimination of All Forms of Discrimination Against Women [CEDAW], a UN treaty, that was adopted in 1979 and came into force in 1981, and the 1995 Beijing Platform for Action, which stresses the need for women to access, own and control natural resources.

Notes

1 The French feminist writer Françoise d'Eaubonne coined the term "ecofeminism" in 1974. There are different forms of feminism and ecofeminism.

2 The *Siiqqee* institution of the Oromo women in Ethiopia is another important example (see below).

3 For instance, women have been able to own and control resources and participate in livestock and land management in Zimbabwe after independence. "Women have also begun to access input loan schemes and to register property including livestock in their own names" (Mark and Winniefridah, 2013, p. 336). The political participation of women in South Africa, Rwanda and Uganda is getting better over the years (see Gordon, 2013, pp. 316–317). Although the South African constitution of 1996 established the Commission for Gender Equality and some women became the members of the parliament, South African women have continued to be the victims of rape and domestic violence (Nkenkana, 2015). After the genocide, women accounted for 70 per cent of the Rwandan population, as some men passed away, others have been imprisoned and left the country. Accordingly, the demographic structure of the country after the

genocide helped women to take positions in the formal sector (Nkenkana, 2015, p. 54). Ethiopia, Ghana, Zimbabwe, Cote d'Ivoire, Senegal, and Kenya have made reforms to address gender issues in their respective countries since the mid-1980s (Gordon, 2013, pp. 325–326).

4 In fact, some men do not allow women to hunt wildlife and possess the knowledge of medicinal plants in Guinea and Southwest Ethiopia in the belief that women cannot keep secrets and might reveal the location of secret hunting grounds (Leach, 1999, cited in Flintan and Tedla, 2010, p. 6) and secret recipes respectively (Flintan, 2004, cited in Flintan and Tedla, 2010, p. 6).

5 Similarly, Wane and Chandler (2002) showed that the Embu women in Kenya have the knowledge of their environment, as they are connected to the land and the environment.

6 Men can also milk cows in Zimbabwe. Sometimes there are no rigid masculinised and feminised roles in Africa,

7 In most cases, women, for instance Afar women in Ethiopia (Balehegn and Tafere, 2013), are not encouraged to move from place to place and engage in petty trade.

8 The Japanese people have a similar proverb: "A woman's mind and autumn sky [change often]" (Schipper, 2006, p. 295).

9 Similarly, African feminists criticised *ubuntu* (humanness) for being the root of female oppression in *ubuntu* societies in Africa. They are of the opinion that *ubuntu* as a patriarchal philosophy is responsible for the violation of the human rights, dignity and equality of women, as in *ubuntu* philosophy women and men are not equal. In particular, they stress that the fundamental African value of hospitality, virginity testing and female genital mutilation (FGM) violate the human rights of African females (for details, see Keevy, 2008). It should be noted that other writers praise the value of *ubuntu* philosophy to Africa and the world, because they think that it is against unethical behaviours and the suffering of fellow human beings (see Mulemfo, 2000). A detailed discussion of arguments for and against *ubuntu* philosophy is beyond the scope of this paper.

10 Likewise, Yoruba men are required to protect woman as the weaker sex (Idowu, 1982, p. 164, cited in Familusi, 2012, p. 307).

References

Aidoo, A. A. (1998). The African Woman Today. In O. Nnaemeka (Ed.), *Sisterhood, Feminisms and Power: From Africa to the Diaspora* (pp. 39–50). Trento, NJ and Asmara: Africa World Press, Inc.

Ajibola, J. O. (2005). *Owe Yoruba*. Ibadan: Ibadan University Press.

Amadiume, I. (1987 [2015]). *Male Daughters, Female Husbands: Gender and Sex in an African Society*. London: Zed Books.

Amsalu, A., and Wana, D. (2013). Climate Change Impacts and Local Coping Strategies Among Pastoral Women: Cases From the Southern Lowlands of Ethiopia. In M. M. Mulinge and M. Getu (Eds.), *Impacts of Climate Change and Variability on Pastoral Women in Sub-Saharan Africa* (pp. 17–44). Kampala and Addis Ababa: Fountain Publishers and OSSREA.

Arnfred, S. (Ed.) (2004). *Re-Thinking Sexualities in Africa*. Uppsala: Nordiska Afrikainstitute.

Balehegn, M., and Tafere, K. (2013). Gendered Impacts and Adaptation Mechanisms to Climate Change Among Afar Pastoralists in North Eastern Ethiopia. In M. M. Mulinge and M. Getu (Eds.), *Impacts of Climate Change and Variability on Pastoral Women in Sub-Saharan Africa* (pp. 83–119). Kampala and Addis Ababa: Fountain Publishers and OSSREA.

Balogun, T. (2005). A Semantic Enquiry Into Yoruba Proverbs and Gender Discrimination. *Journal of Arts and Humanities*, 2(1), 39–43.

Bartels, L. (1983). *Oromo Religion: Myths and Rites of the Western Oromo of Ethiopia. An Attempt to Understand*. Berlin: Dietrich Reinner.

Blystad, A. (1995). Peril or Penalty: AIDS in the Context of Social Change Among the Barabaig. In K-I. Klepp, P. M. Biswalo and A. Talle (Eds.), *Young People at Risk* (pp. 86–105). Oslo: Scandinavian University Press.

Coquery-Vidrovitch, C. (1997). *African Women: A Modern History*. B. G. Raps (Trans.). Boulder, CO and Oxford: Westview Press.

Debsu, D. N. (2009). Gender and Culture in Southern Ethiopia: An Ethnographic Analysis of Guji-Oromo Women's Customary Rights. *African Studies Monographs*, 30(1), 15–36.

Etefa, T. (2012). *Integration and Peace in East Africa: A History of the Oromo Nation*. New York: Palgrave Macmillan.

Familusi, O. O. (2012). African Culture and the Status of Women: The Yoruba Example. *The Journal of Pan African Studies*, 5(1), 299–313.

Flintan, F., and Tedla, S. (2010). Introduction. In F. Flintan and S. Tedla (Eds.), *Natural Resource Management: The Impact of Gender and Social Issues* (Revised ed., pp. 1–18). Kampala: Fountain Publishers.

Gordon, A. A. (2013). Women and Development. In A. A. Gordon and D. L. Gordon (Eds.), *Understanding Contemporary Africa* (pp. 305–333). Boulder, CO and London: Lynne Rienner Publishers.

Hussein, J. W. (2004). A Cultural Representation of Women in the Oromo Society. *African Study Monographs*, 25(3), 103–147.

Jacobson, L. J. (1993). Closing the Gender Gap in Development. In L. R. Brown (Ed.), *State of the World* (pp. 61–79). New York: W. W. Norton.

Kalu, O. U. (2001). The Sacred Egg: Worldview, Ecology, and Development in West Africa. In J. A. Grim (Ed.), *Indigenous Traditions and Ecology: The Interbeing of Cosmology and Community* (pp. 225–248). Cambridge, MA: Harvard University Press.

Keevy, I. (2008). *African Philosophical Values and Constitutionalism: A Feminist Perspective on Ubuntu as a Constitutional Value*. Unpublished Doctoral Thesis, Faculty of Law, University of the Free State.

Kelbessa, W. (2001). Traditional Oromo Attitudes Towards the Environment: An Argument for Environmentally Sound Development. *OSSREA Social Science Research Report Series, No. 19*, Organisation for Social Science Research in Eastern and Southern Africa, Addis Ababa.

Kelbessa, W. (2011). *Indigenous and Modern Environmental Ethics: A Study of the Indigenous Oromo Environmental Ethic and Modern Issues of Environment and Development: Ethiopian Philosophical Studies. I*. Washington, DC: The Council for Research in Values and Philosophy.

Kiiru, M. (1999). You Cannot Catch Old Birds With Haff: Woman's Multiple Images in Proverbs. *WAJIBU: A Journal of Social & Religious Concern*, 14(1). Available at http://web.peacelink.it/wajibu/6_issue/p3.html

Kipuri, N., and Ridgewell, A. (2008). *A Double Bind: The Exclusion of Pastoralist Women in the East and Horn of Africa*. London: Minority Rights Group International (MRG).

Kumsa, K. (1997). The Siiqqe Institution of Oromo Women. *The Journal of Oromo Studies*, 4(1/2), 115–152.

Legesse, A. (2000). *Oromo Democracy: An Indigenous African Political System*. Lawrenceville, NJ: The Red Sea Press.

Maathai, W. (2007). *Unbowed: A Memoir*. New York: Anchor/Random House.

Maathai, W. (2010). *Replenishing the Earth: Spiritual Values for Healing Ourselves and the World*. New York: Doubleday.

Mark, M., and Winniefridah, M. (2013). Traditional Adaptation Mechanisms to Climate Change and Variability Among Women Pastoralists in South-West Zimbabwe. In M. M. Mulinge and M. Getu (Eds.), *Impacts of Climate Change and Variability on Pastoral Women in Sub-Saharan Africa* (pp. 297–342). Kampala and Addis Ababa: Fountain Publishers and OSSREA.

Mawaya, C., and Kalindekafe, M. P. (2010). Access, Control and Use of Natural Resources in Southern Malawi: A Gender Analysis. In F. Flintan and S. Tedla (Eds.), *Natural Resource Management: The Impact of Gender and Social Issues* (Revised ed., pp. 88–124). Kampala: Fountain Publishers.

Mulemfo, M. M. (2000). *Thabo Mbeki and the African Renaissance*. Pretoria: Actua Press.

Mulinge, M. M. (2013). Introduction. In M. M. Mulinge and M. Getu (Eds.), *Impacts of Climate Change and Variability on Pastoral Women in Sub-Saharan Africa* (pp. 1–15). Kampala and Addis Ababa: Fountain Publishers and OSSREA.

Nabalamba, A., Mubila, M., and Alexander, P. (2011). *Climate Change, Gender and Development in Africa*. Abuja: Africa Development Bank 1(1).

Njiru, B. (2013). Challenged Livelihoods as a Result of Water Scarcity Among Maasai Women Pastoralists in Kajiado County, Kenya. In M. M. Mulinge and M. Getu (Eds.), *Impacts of Climate Change and Variability on Pastoral Women in Sub-Saharan Africa* (pp. 45–81). Kampala and Addis Ababa: Fountain Publishers and OSSREA.

Nkenkana, Akhona. (2015). No African Futures without the Liberation of Women. *Africa Development* 3, 41–57.

Nzegwu, N. (2006). *Family Matters: Feminist Concepts in African Philosophy of Culture*. Albany: Suny Press.

Olsson, A. I. (1994). Somali Women in Reconciliation and Ecological Recovery. In M. A. M. Salih and L. Wohlgemuth (Eds.), *Crisis Management and the Politics of Reconciliation in Somalia* (pp. 73–75). Uppsala: Nordiska Afrikainstitute.

Omari, C. K. (1990). Traditional African Land Ethics. In J. R. Engel and J. G. Engel (Eds.), *Ethics of Environment and Development: Global Challenge, International Response* (pp. 167–175). Tucson, AZ: University of Arizona Press.

Presbey, G. M. (2013). Women's Empowerment: The Insights of Wangari Maathai. *Journal of Global Ethics*, 9(3), 277–292.

Schipper, M. (2006). *Never Marry a Woman With Big Feet*. Amsterdam: Amsterdam University Press.

Seager, J. (2009). *The Penguin Atlas of Women in the World* (4th ed.). New York: Penguin Books.

Shiva, V. (1995). *Staying Alive: Women, Ecology and Survival in India*. New Delhi: Indraprastha.

Shongolo, A. A. (2004). Boran Traditional Blessings and Prayers. In C. Griefenow-Mewis and T. Bitima (Eds.), *Oromo Oral Poetry Seen From Within* (pp. 49–57). Koln: Rüdiger Köppe Verlag.

Sikhondze, B. (2003). The Role of Women in Food Production in Swaziland, 1945–65. In B. Zewde (Ed.), *Land, Gender and the Periphery: Themes in the History of Eastern and Southern Africa* (pp. 67–79). Addis Ababa: Organisation for Social Science Research in Eastern and Southern Africa.

Sofola, Z. (1998). Feminism and African Womanhood. In O. Nnaemeka (Eds.), *Sisterhood, Feminisms and Power: From Africa to the Diaspora* (pp. 51–64). Asmara and Trenton, NJ: Africa World Press.

TAVANA. ND. The Tree Mother of Africa and Her Green Belt Movement. Available from https://tavaana.org/en/content/tree-mother-africa-and-her-green-belt-movement-1

Toit, L. D. (2011). Old Wives' Tales and Philosophical Delusions: On 'the Problem of Women and African Philosophy'. In G. Walmsley (Ed.), *African Philosophy and the Future of Afric*a (pp. 111–128). Washington, DC: The Council for Research in Values and Philosophy.

Wako, T. N. (2011). *In Praise of Saayyaa: Oromo Songs in Praise of the Cattle*. Harare: PCV Publishers.

Wane, N., and Chandler, D. J. (2002). African Women, Cultural Knowledge, and Environmental Education With a Focus on Kenya's Indigenous Women. *Canadian Journal of Environmental Education*, 7(1), 86–98.

8 Women's identities in African environmental ethics

A conversational approach

Ademola Kazeem Fayemi and Oladele Abiodun Balogun

Introduction

Environmental ethics is one of the emerging areas of concern in contemporary African philosophy. In the conservatively five decades of postcolonial history of African philosophy, very few voices of women professional philosophers have been heard. "As a loose set of canonical texts and authors," African philosophy is "currently and overwhelmingly dominated by men" (du Toit, 2008, p. 413). While the underrepresentation of women in general in the African philosophical space may be an unintended case of marginalisation, in particular, there are relatively few but significant women authors in African environmental discourses. Amidst the absence of a demographical database on women in philosophy in Sub- Saharan Africa, this chapter aims to discuss women's identities and presence in African environmental ethics. In this regard, the views of Sophie Oluwole, Philomena Ojomo and Pat Agboro, who are all Nigerian philosophers, shall be discussed.

These scholars explore, in different ways, the thesis of indigenous African environmental values and principles, which are considered sacrosanct in the restoration of humans and the environment in Africa. However, no serious critical investigations have been carried out on the nexus, strengths and flaws in the emerging theoretical views on African environmental ethics. This chapter draws upon the view of Wangari Maathai, a Kenyan, in challenging and corroborating the views of Oluwole, Ojomo, and Agboro. Essentially, this article aims precisely at further interrogating the thoughts of these four women scholars on African environmental ethics through the conversational approach.

The conversational approach is a "philosophical engagement between individual thinkers with one another; on phenomenological [existential and substantive] issues of concern; or on one another's thoughts where thoughts are unfolded from concepts, or from concept of concepts" (Chimakonam, 2015, p. 20). Beyond being dialogic engagement between or among scholars, the conversational approach contests and protests existing ideas and thoughts to create new concepts; it is not restrictively bound in African philosophical space and place; it is a growing "practice of philosophy in our age generally" (Chimakonam, 2015, p. 20).

The attempt to conversationally engage the thoughts of African women subjects in environmental ethical discourses in Sub-Saharan Africa is historically important in order to: (i) understand the extent of the marginalisation of women in the existing discourse; (ii) provide an overview of the trajectory of African environmental ethics; (iii) further the discourse beyond theoretical polemics to more substantive issues on women and environmental management in a way that shows the direct implications of thoughts for praxis.

African environmental ethics and the loud voices of men

The impression that African philosophy is a male-dominated discipline is strong. Sanya Osha echoes this when he avers that "modern African philosophy is a product of a patriarchal culture and hence a sexist one" (2006, p. 157). As a matter of deduction, given that African environmental ethics is a subset of African philosophy, which is patriarchally dominated, the inference can be made that there is marginalisation and silencing of women in African environmental ethics. However, the cogency of this claim will depend on the truth of the male-sexist character of African environmental ethics and the identity of its authors. A survey of the scholars that have contributed to the budding literature on African environmental ethics can shed light on the extent of the marginalisation of the women in the existing discourse. Though the list is exhaustive, identified for specific discussion here are male African philosophers like Segun Ogungbemi, Godfrey Tangwa, Ayo Fadahunsi, Chigbo Ekwealo and Workineh Kelbessa. While the views and postulations of these scholars on African environmental ethics are exposed, shortly, it does not necessarily follow that African environmental ethics is systematically designed to be male-oriented and dominated.

Approaching environmental problems from the ethical angle has a fundamental presupposition. This is that environmental crises are largely the result of human actions. Given that axiological systems inform actions, it is logical to query the ethical underpinning of human actions, relations and activities with the environment. In line with this thinking, African environmental ethics is therefore defined by Ekwealo as "an applied philosophy dealing with the fundamental principles that govern the relationship between man and the environment based on African worldview" (Ekwealo, 2011, p. 4). Interest is on the increase by African scholars in enunciating the effectiveness of an environmental ethics that will be within the gamut of African values and culture. They explicate African environmental philosophical theories which they thought (if adhered to strictly) would address the African environmental problem. Africa is heterogeneous in identity and other cultural forms, and African environmental ethics is not a monolithic thought framed in the garb of uniqueness of theories and principles. Rather, it is the critical reflections on the "underlying common or shared environmental values of African societies" (Kelbessa, 2015, p. 389).

Ogungbemi (1997) puts forward an 'ethics of care.' By this, he means an orientation in which one is not taking more than one's needs from nature. He notes that the management of the ecosystem, valuing and respecting nature is not something alien to the Africans. According to him:

> In our traditional relationship with nature, men and women recognise the importance of water, land and air management. To our traditional

> communities the ethics of not taking more than you need from nature is a moral code. Perhaps this explains why earth, forests, rivers and wind and other natural objects are traditionally believed to be both natural and divine. The philosophy behind this belief may not necessarily be religious, but a natural means by which the human environment can be preserved. The ethics of care is essential to traditional understanding of environmental protection and conservation.
>
> (Ogungbemi, 1997, p. 204)

Ogungbemi, however, reformulates the traditional orientation on the environmental practice of an 'ethics of care' to make it fit into the contemporary African situation, and this he calls 'ethics of natural relatedness,' which he explains thus:

> Ethics of nature relatedness asserts that our natural resources do not need man for their existence and functions . . . the ethics of nature relatedness can be succinctly stated as an ethics that leads human being to seek to co-exist peacefully with nature and treat it with some reasonable concern for its worth, survival and sustainability.
>
> (Ogungbemi, 1997, pp. 208–209)

He further pontificates three practical ways to halt the current environmental mismanagement in Africa. First, solar energy should replace all other forms of energy generation. Second, population needs to be controlled, as overpopulation leads to natural disasters. Lastly, a reasonable government should be committed to friendly environmental policies, which will cut the excesses of industrialisation and agricultural wastes by disposing of them properly (Ogungbemi, 1997, p. 209).

Ogungbemi's ethical postulation on the African environment and his practical recommendations are not without flaws. Critically, Ogungbemi's ethics of nature-relatedness has no link with African cultural experience; nor is it grounded in any known African ontological scheme. In arguing against his first option on the use of solar energy, Babatunde Fashola (2008, p. 16) points out that, as a possibility, there is nothing provided by nature that is inexhaustible. Fashola holds that there is the need for us to ascertain how exhaustible or inexhaustible the solar system itself is before we pursue recourse to it as an alternative source of energy. While not oblivious of the pollution caused by gasoline consumption, which leads to global warming, Fashola urges caution, rationality and verification in exploring solar alternatives so as to avoid replacing one problem with another. He therefore posits that the way out of the African environmental problem is to embark on a scientific search for alternative energy resources and waste management. While, in truth, his order of priority is not entirely misplaced, it is worth noting that the solution to environmental destabilisation is not purely technological or exclusively attitudinal. There are ethical concerns and values and principles that will guide even such technological alternatives.

Seeing science, technology and civilisation and their underlying values as having contributed to the continuous deterioration of the African environment, Fadahunsi, for his part, opted for a holistic conception of environmental ethics. This involves all strands of environmental ethics and requires that attention be paid to the wisdom and limitations of each, since none of them, considered alone, can sufficiently deal with the environmental crisis (Fadahunsi, 2007, p. 10). He dubs this ethical theory 'holistic environmental ethics.' Fadahunsi's position is arrived at through his critical examination of the flaws and merits in the various existing strands of environmental ethics in Western intellectual tradition: anthropocentricism, animal liberation/rights theory, biocentrism and ecocentrism. He argues that 'holistic environmental ethics' is a synergy and integration of the positive aspects of the existing environmental ethical theories in Western scholarship. Holistic environmental ethics, according to Fadahunsi, expresses four moral duties – our duty to humans, our duty to animals, our duty to all life forms and our duty to nature as a whole – which are essential in protecting the world environment from further deterioration (Fadahunsi, 2007, p. 11).

Whatever the substance of 'holistic environmental ethics,' as presented by Fadahunsi, it is doubtful whether it can be rightly called an African environmental ethical theory; it is more of an environmental ethical theory by an African. The Africanness of the theory is seriously in doubt; neither is it a reflection on indigenous African ethical paradigms and values or a concerted endogenous adaptation of non-African environmental ethics to the African experience of environmental problems. Much like its predecessors in Western scholarship that ended up at the level of theory, without practical template, 'holistic environmental ethics' is unlikely to be the antidote to the African condition.

Tangwa came up with what he calls 'eco-bio-communitarianism' as an African ethical theory. 'Eco-bio-communitarianism' is based on the metaphysics of traditional African societies, which, according to him, involves the "recognition and acceptance of inter-dependence and peaceful coexistence between earth, plants, animals and humans" (Tangwa, 2004, p. 389). Given this metaphysical disposition of the traditional Africans, Tangwa says that traditional Africans were more cautious in their attitude to plants, animal and inanimate things and the various invisible forces of the world (Tangwa, 2004, p. 389). Such a metaphysical understanding allows that the dichotomy between "plants, animals, and inanimate things, the sacred and the profane, matter and spirit, the communal and the individual, is a slim and flexible one" (Tangwa, 2004, p. 389). Tangwa notes that traditional African societies were communitarian in structure, practice and relation to the environment; their attitude towards nature and the rest of creation is that of respectful coexistence, conciliation, and containment. He suggests, therefore, an eco-bio-communitarian African environmental ethics which will serve as the basis of contemporary Africans' interaction with the environment. Such an ethics entails a live and let-live attitude, which is sacrosanct to the restoration of the African environment.

Tangwa's exploration of an African environmental ethics has been criticised by Ojomo (2011, p. 110), who noted that the attempt is more of an ethnophilosophical defense of indigenous African treatment and management of the environment. This is because Tangwa never recognised the ways and manners by which traditional Africans contributed to the degradation of the environment, through ignorance, poverty and iconoclastically motivated cultural religious practices that devalue the integrity of other sentient beings that humans share the environment with. Moreover, the supposed communalistic conception of traditional African life is questionable, given that many thinkers have argued against it in recent literature.

The view of Ekwealo on African environmental ethics is worthy of consideration. He defends what he calls 'ecocentric' African environmental ethics, which he strongly believes is markedly different from the Western option. His argument is that "African environmental philosophy has a background belief in the linkage of nature, community and man, from which an ethical relationship is defined" (Ekwealo, 2011, p. 4). The ecocentric nature of African environmental ethics, according to Ekwealo, is against the backdrop of the African people's culture of according other lives and creatures that exist some modicum of respect, dignity and recognising the 'rights' or 'value' that all beings possessed (Ekwealo, 2011, p. 5). Ecocentricism ensures the utilisation of adequate resources from nature for human efficient use, while at the same time ensuring qualitative life for the environment itself.

Ekwealo explores the metaphysical underpinning of African environmental philosophy. He contrasts African metaphysical dualism with the Western conception, according to which objects of dualist theory exist in competition. In African metaphysical dualism, he maintains, "everything has a complementary opposite, and the interaction of the two produces good" (Ekwealo, 2011, p. 5). In this metaphysics, there is an awareness that everything exists to help in the realisation of the other; this connection accounts for the harmonious relationship between man and the other constituents of the environment. In Ekwealo's analysis, Western environmental philosophy is different from African environmental philosophy in a number of ways: whereas in Western environmental philosophy, life is restricted to animate and human beings and there is a distinction between man and the environment, in African environmental philosophy, there is a life-force in nature and, because of this, both man and environment are traced to the primordial force.

The interrelation among nature's life forces leads to the commonplace "conservationist moral values and attitudes among Africans" (Ikuenobe, 2014, p. 1). Polycarp Ikuenobe, in consonance with Ekwealo, recognises that forces (divine, celestial or terrestrial, human, animal, vegetal, material or mineral) exist in unity and interact in order to achieve harmony, respect and conservation of nature. Both agree that, unlike in the West, where the survival of the fittest syndrome explains man's evolutionary history, man in African environmental philosophy is a member of the ecosystem. The distinction between environmental philosophy in the Western and African contexts is therefore premised on this. To quote Ekwealo (2011, p. 6): "Western environmental philosophy is monistic whereas African environmental philosophy is pluralistic." This pluralism, according to him, necessitates the ethical principle on which humans' relationship with the environment is based: respect, accommodation and tolerance of other entities that the human shares the environment with. Following his ecocentric basis of African environmental philosophy, Ekwealo advocates spiritualism of the environment as an imperative for the restoration and protection of the environment. A spiritualised environment by ecological beings will lead to "a renunciation of excessive worldly goods, a promotion of frugality against consumerism" (Ekwealo, 2011, p. 9), which are the root of man's domination of the environment.

Critically, Ekwealo's attempt at an African environmental philosophical theory has some flaws. He based his analysis of African environmental ethics on his understanding of African metaphysics, which he calls dualism. What is problematic here is not the explanation given to this metaphysical system; rather, it is the interpretative name ascribed to it – 'dualism.' The metaphysical structure of the African nature of reality falls within the binary conceptual scheme and can best be rationally comprehended when located within it. To say, as Ekwealo does, that it is dualism of a certain kind different from what obtains in the West is to conflate the same category as otherness. In this new interpretation, the physical and spiritual constituents of reality are not separate entities (as we have in dualism) but rather one entity in two interdependent folds; both are two sides of the same coin.

Besides this point, Ekwealo regards African environmental ethics as ecocentrist in orientation and argues that it is different from what obtains in Western environmental philosophy. But his explanations, essentially, do not reflect that what ecocentrism means for Africans differs from the Western understanding of it. Ecocentrism focuses on the integrity of the ecosystem and the value of species (animate, human, among others) within it. In Western environmental philosophical literature on ecocentrism, this orientation is not even presented as a monolithic viewpoint; we have the land ethic, deep ecology and the theory of nature's value. The question is, in which of these variants does the African ecocentric environmental theory belong, or how is African environmental ecocentric theory different from them? Ekwealo's argument would have been more enlightening and convincing if he had considered such fundamental questions. Whatever the stance one takes, ecocentricism itself is questionable because it requires excessive dutiful sacrifice by humans in a bid to ensure ecological balance.

More problematic is Ekwealo's comparative stance that Western environmental philosophy is monistic, while that of Africa is pluralistic. Given the diversity of ethical theories in Western environmental philosophy, it is misleading to depict Western environmental philosophy as monistic. Moreover, to accept the spiritualisation of the African environment, as Ekwealo has advocated, leads to grave implications. We may ask, what becomes the duty of secularly minded agents in Africa to the environment? Must one be spiritually inclined to be environmentally conscious? Indeed, if traditional Africans saw the environment as sacred, this was more of an unintended consequence and not intentionally motivated; it was a consequence of the absence of industrialisation. A reinvigoration of this inglorious past in this age is, after all, retrogressive.

From the foregoing reflective attempts by male African philosophers to discover the principles underlying environmental consciousness in traditional African cultures and to formulate principles that can mitigate environmental problems in contemporary Africa, some pertinent questions can be raised. To what extent do women contribute to environmental ethical discourse and environmental advocacy in sub-Saharan Africa? Do sub-Saharan African worldviews accommodate the subordination of women in environmental matters in particular? Does African environmental ethics marginalise women in its scope of reflection? Does 'ecological feminism' (or 'ecological womanism') influence or inspire the writings of women authors in African environmental ethics, or have their works been conditioned by some other salient identity strictures?

Women and the African environment: Sophie Oluwole's perspective

Oluwole is one of Africa's foremost philosophers and a distinct female African philosopher, who works extensively in the area of oral tradition as philosophy and has made original and uniquely perceptible contributions to the exposition of ancient Yoruba philosophy and African philosophy in general. While she has written copiously on the question of women and gender relations in Yoruba culture with a less comprehensive treatise on African environmental ethics, her major metaphysical thesis is salient to her projected stance in African environmental ethics. Her metaphysical thesis is binary complementarity, which is the position that matter and non-matter exist interdependently in a way that cannot be separated (Oluwole, 2014). The Yoruba metaphysical worldview is centered on binary complementarity, a view that all existences as diverse as they are – environment, human beings expressed in sexist terms of maleness and femaleness, politics, religion, morality, among others – cannot operate independently without interconnectedness involving necessarily, inclusionary paradoxes and contradictions.

At the 2012 international conference on African environmental ethics organised by the African Environmental Ethics and Values Research Group, Department of Philosophy, University of Lagos, in collaboration with the Center for Environmental Philosophy, University of North Texas, Oluwole in her guest lecture defends a metaphysical basis for any future African environmental ethics, which will be based on binary complementarity. This binary metaphysical axiom involves a necessary respect, tolerance and inalienable separation of human beings from nature and the environment as a whole. Binary complementarity connects with the debate on the marginalisation of women in different strata of existence and the advocacy of equality. Binary complementarity also resonates with the intellectual search for a viable principle of African environmental ethics. Coming to grasp with her concept of binary complementarity is apt to underscore and understand its implications for the question of women and the environment.

Oluwole, in establishing a ground for binary complementarity, writes:

> Earliest African thinkers did not conceptualise nature as one monolithic reality or as two oppositional realities that can never blend or be reconciled. For them, matter and mind – the latter expressed as spirituality and not as religion – was conceptualised as one reality with binary features, two inseparable phenomenal charges as the atom is today defined in nuclear physics. The symbolism adopted to give this a concrete image is the coin. If you separate

> the two sides you do not have a coin just as you do not have an atom when you remove either the positive or negative charges. I classify this metaphysics as binary complementarity.
>
> (Oluwole, 2011, p. 14)

The above African binary metaphysical principle, as explicated by Oluwole, has some epistemological, axiological, socio-political, economical, logical, ethical, environmental and gender principles derivable from it. At the level of epistemology, experience guided by reason will give reliable knowledge and not absolute knowledge. Axiologically, it implies that there is value in that which seems valueless. Socio-politically, "I/others" constitutes an inseparable social unit. Economically, binary complementarity implies social welfarism; jurisprudentially, it means that every citizen has rights to get the best out of life without necessarily using others as mere economic stooges. Logically, 'either . . . or' without aversion for contradiction becomes the ordering mental principle of evaluating thought. Ethically, morality is a matter of sympathetic and caring concern of the humanity of others and the recognition of the being of others (Fayemi, 2013).

Environmentally, the binary complementarity principle implies that all forms of relationship in the environment are identified as ecologically complementary to one another. "Binary complementarity as a social metaphor is an expression of an inalienable respect for human beings as well as for our environment" (Oluwole, 2012). Though Oluwole's inquisition in African environmental ethics does not yet detail the nature of environmental problems in Africa and how it can be managed, there is a sense in which her works seem to answer the fundamental question of whether or not sub-Saharan African worldviews accommodate the subordination of women in environmental matters in particular and in other spheres of institutional existence.

Two of her works that show the above concern are: "Culture, Gender and Development Theories in Africa" (1997) and her co-authored book, with J. A. Sofoluwe, *African Myths and Legends of Gender* (2012b). In these works, Oluwole employed the approach of hermeneutic interpretation of the social history of the Yoruba as expressed in proverbs, aphorisms, fables and myths in arguing that malignant sexism is repugnant in the Yoruba worldview and that external factors are more apt in explaining the marginalisation of women in African societies, traditional or contemporary. While natural physiological differences "between the male and female cannot be denied, nature does not by itself justify male-female absolute equality. But neither are these differences synonymous with women's total inferiority" (Oluwole, 1997, p. 112) as both sexes are of the same species. Oluwole's point is that, just as with the Yoruba, many indigenous sub-Saharan traditional African societies did not intentionally deny or deter women from performing creditable roles in spiritual, intellectual, economic, political and environmental matters.

In *African Myths and Legends of Gender*, Oluwole surveys some of the myths created by the indigenous Yoruba to explain the roles of male and female in human society. According to her, "African heroines and heroes are imageries drawn from some historical experiences, in which some women made undeniable contributions to the development of their societies" (Oluwole and Sofoluwe, 2012, p. 16). In many of the oral traditions she explored, historical evidence abounds of women achieving a revered heroic status, becoming, for example, monarchs, war mongers, singers, medicinal experts, saviours, poets, artists, pathfinders and environmental conservationists. Such myths showed that for the Yoruba, "the involvement of women in all social institutions is a divine order (natural not spiritual), one of God's commandments that must be respected and implemented, if society is to be peaceful and be in a state of harmony" (Oluwole, 2012).

Despite the feats recorded by the women folk, male chauvinism and marginalisation were never totally absent at the intellectual, social, political, environmental management and other spheres in Africa. This explains the paradoxes and apparent contradictions accommodated within the metaphysical principle of binary complementarity in the Yoruba worldview. The salient point in the binary complementary construct of Oluwole in relation to the question of the gender gap, whether in philosophical circles, environmental consciousness or other human endeavours, is that "a woman can be strong or weak; she can be gentle or harsh; she can be moral or immoral just as her male counterparts can equally be" in the exercise of capabilities in any area of human endeavour (Oluwole, 2014 p. 201). What is most crucial is the recognition of male-female complementarity in creating a healthy environment for the flourishing of other human capabilities. Hence the Yoruba proverb: "*bi oôkunrin r'ejo, t'obinrin pa a; ki ejo ma sa a ti lo*" (If a man sights a snake and a woman kills it, all is well, so long the snake does not escape).

V. Y. Mudimbe's view seems to be in tandem with Oluwole on the position that sub-Saharan African cosmological worldviews do not inspire the subordination of women in environmental matters. While Oluwole uses Yoruba-African mythology and proverbs as examples to make her point, Mudimbe interrogates the figure of woman in parables and fables according to Luba myths and cosmology in coming to a similar conclusion as did Oluwole. In his "Creation, History and the Sex of Beings," Mudimbe demonstrates how the elements – air, wind, water, forests, birds, trees, and animals and earth – are all mediated by dynamics of gender. Beyond wo/man, gender exists giving the universe its myriad forms and investing the rhythms of life with its flows, transformations (metanoia) and its pulse (Mudimbe, p. 75). Unlike Oluwole, who emphasises a metaphysical axiom, Mudimbe provided an analysis of "hermaphroditism not only as a cosmological principle but by contributing to its doubling and its conceptual possibilities" (Osha, 2006, p. 179). Lacking in both is a practical template and roadmap for restoring environmental health, protecting environmental integrity for current and the future generations.

African environmental ethics: Philomena Ojomo's perspective

Let us now turn to the view of Ojomo, a Nigerian female environmental philosopher. Her concern is different from Oluwole, who seeks to provide a metaphysical grounding for African environmental ethics. Nor is she motivated by criticising existing African environmental ethical theory defended by male African philosophers. Wary of not overpopulating the existing attempts at theorising on African environmental ethical theory, Ojomo in her "Environmental Ethics: An African Understanding," states that "environmental ethics, no matter how grounded in African experience it could be, or intellectually sophisticated it could be, can alone solve the environmental crisis in Africa" (Ojomo, 2011, p. 111). She is of the view that there has to be an orientation that keeps in mind other non-philosophical ideas and recommendations. In her words:

> The task of fashioning what should constitute the ethical principles of the African environment is one of the most fundamental challenges and issues contemporary African philosophy continent must deal with. Not until this intellectual imperative is settled, collaborative efforts by government, NGOs, environmental agencies and mangers, as well as other sub sectors of the society towards orientation and sensitization, policy formulation and implementation on sustainable environment in Africa would remain a pull in the dark. (Ojomo, 2011, p. 113)

Inspired by the concern for seeking solutions, Ojomo argues that the environmental crises in Africa can be dealt with if both government and non-governmental bodies contribute their quota in ensuring a good environment in Africa through public enlightenment, environmental legislation and enforcement, among other things, on the need to preserve and protect the environment. In her article "Aldo Leopold's Land Ethic: Implications for Africa," Ojomo (2012) argues that achieving such a goal is impossible except when anchored on a socialist frame, which she calls earth-eco-socialism. By this she meant a theoretical reinforcement and synthesis of "a socialist ideological and ecological action in understanding of the intrinsic value of the earth as indispensable platform in the quest for environmental sustainability"(Ojomo, 2012, p. 1).

However, Ojomo does not provide an explicit view on how she thought this enlightenment could be achieved – whether through the media or through community meetings, or environmental activism or through other means. However, she is quick to point out that environmental consciousness should be instilled into the student by introducing environmental issues in their course content. Details and an outline of how this could be done, as well as at what level of education should such environmental tutorials be introduced, are, unfortunately, lacking in her analysis. The socialist strand of her theory makes it neither African nor purely Western in constitution because of the convoluted contemporary African environmental distinctiveness, Marxist and Leopoldian frames in which the theory of earth-eco-socialism is anchored. In seeking an eco-socialist exit of the African environmental crises, the explicit assumption of this theory is that the root of the environmental problems is in the ideology of capitalism. However, given the many-sidedness of the causes of environmental crises in contemporary Africa, Ojomo's socialist ideological hypothesis is susceptible to reductionism, just as it does not also factor the increasing globalising effects and influences of economic and political ideas, and consumption patterns from non-socialist societies on the Africans and their environment.

Externalism and African environmental ethics: Pat Agboro's perspective

Agboro is another female Nigerian philosopher whose article added another dimension to the growing literatures on African ethics by female authors. Her motivation is neither to fashion a metaphysical foundation of African environmental ethics nor to defend the distinctiveness and necessity of African ethical canons in relation to the environment. She is more concerned with seeking the root of the fundamental distortions in African value systems, which have continued to be desecrated in contemporary times with untoward implications on the African environment. Her main thesis is that the external colonial distortions of African values and Africans' psyche are responsible for environmental crises in Africa. The problem of the African environment, in her view, is peculiar:

> The peculiarities are entrenched in colonialism and a resultant distortion of the psyche of the African man. This has reduced him to a mere being struggling for survival . . . As he seeks survival, he consequently destroys that which would have been of benefit to him . . . [consequently leading to] the myopic stance of 'it is better to be rich and dirty than to be poor and clean.'
>
> (Agboro, 2008, p. 5)

Human activities that have necessitated environmental problems in Africa today are products of the colonial structures and philosophy of domination. Prior to colonialism, Agboro argues, traditional Africans considered nature as an end in itself deserving respect and reverence. However, in contemporary Africa, the same attitude is dwindling with the popular colonialist's niche of conquering and survival. A significant difference between pre-colonial Africa and the West is the latter's moral attitudes towards nature, which is more of domination than conservational. Ikuenobe eloquently expresses this point:

> For Europeans, Africa's rain forests and jungles can be destroyed, lumber can be harvested for homes, furniture, and as export to Europe. The land and forests are to be converted into industrial and real estate complexes or into farms and plantations to produce cash crops and export for industrial Europe. The mountains are to be mined indiscriminately for minerals, the rivers dammed for electricity, and the swamps drilled for oil. These economic activities destroyed various habitats for plants and animals, caused massive erosion, and led to the pollution of the air and rivers.
>
> (Ikuenobe, 2014, p. 12)

While criticising such a fast becoming niche in African world, Agboro argues that the environmental crises cannot be addressed in isolation of the rest of the problems weighing down many African states. Such problems, especially those having to do with the stark poverty of the masses, poor quality of live, healthy and social insecurity, are peculiarities occasioned by the historical facts of colonialism, neo-colonialism and the resultant effects on value erosion. Agboro does not mince words in establishing a connection between the poor conditions of many Africans and how such mental, physical and financial conditions constitute a bane by:

> generating further problems for the environment. . . . A man with a polluted mind will necessarily pollute his environment whether consciously or otherwise. This is so because, his values have become distorted. And a distorted value can only lead to a confused state of affairs where one ideology or another is experimented with and discarded at whim.
>
> (Agboro, 2008, p. 10)

As a way out, Agboro makes a case for putting environmental ethics into practice in African societies. She advocates environmental activism grounded in indigenous cultural values and education as the way out of environmental crises in Africa. She argues that environmental values should be instilled as "the foundation of coercive environmental policies" (Agboro, 2008, p. 12).

The explanation by Agboro can be regarded as an externalist account of environmental crises in Africa. Such externalism seeks to trace the fundamental source of environmental problems in Africa to forces, agents and multinational corporations outside Africa. Just like the externalist socio-political explanation of the historical genesis of Africa's underdevelopment as having roots in slave trade, colonialism and neo-colonialism, Agboro's argument establishes a nexus between the environmental and other socio-economic crises of sub-Saharan African societies. Many contradictions are necessarily involved in her exposition of the roots of African environmental problems and the solutions she sought. For one, if the problems are externally generated, common sense dictates that the solution should also be externally designed. While doing so is logical, though with the grave implication of perpetuating the root of the problem, Agboro fails to provide the direction for solving the externally generated problem of the African environment. For another, none of the indigenous values that colonialism supposedly distorted were identified by Agboro. Her view on the linkage of poverty to environmental degradation raises some scepticism, because it embeds the assumption that there are no environmental problems in affluent nations. While this is far from the truth, as historic experiences in advanced technological first world nations have continued to show, also worthy of objection is her gender insensitivity of femaleness in environmental issues. Throughout her discourse, she begs the question of the extent to which she understands the invaluable roles of women in environmental sustainability.

Women and African environment: Prolegomena to future African ethics

So far, the perspectives of three selected female Nigerian philosophers on the environment in Africa and environmental ethics have been discussed and engaged conversationally. It is apposite to mention that the gender gap in African philosophy generally, and in African environmental ethics in particular, can be understood in different ways. Two such ways are considered in this chapter. One is to examine the presence or marginalisation of women in African environmental ethics scholarship. Two is to interrogate the place, connection and role of women in environmental issues. Both are connected in some ways. In respect to the first concern, in the absence of an existing empirical research on the representation and statistical proportion of women in the scholarship on African philosophy in the sub-Saharan space and place, hypothesising on the degree of marginalisation will be premature. The marginal muzzling or underrepresentation of female voices in African environmental ethics discourses is, to say the least, unsystematically premeditated; however, such supposed marginalisation tends to square within the larger patriarchal culture of many sub-Saharan African societies. Rather than promoting, further, the unintended underrepresentation of women in African environmental discourses, a veritable option is to loud the voices, works and activism of African women environmentalists and scholars through the conversational approach. For this reason, this chapter has considered selectively conversing with some of the works on African ethics with a focus on those by Nigerian women.

From our discussions above, the common identities that straddle across the works of these three philosophers are female authorship and the geographical criterion of Africanness of thought. Neither of the positions of the three selected scholars shares a feminist bias for environmental concerns, nor underscores an appreciation of the roles and constraints of women in sustainable environmental management, be it in Africa or elsewhere. While the respective intellectual focuses of the women philosophers examined in this chapter are legitimate, as each author is entitled to her areas of interest, it is equally apposite to underscore the place, connection and role of women in both African environmental ethical discourse and in general promotion of sustainable environment. While the former is an academic pursuit, for the sake of theory, the latter is a matter of day-to-day living without necessarily being academic.

Wangari Maathai, the renowned Kenyan woman that founded the Green Belt Movement and a recipient of the 2004 Nobel Peace Prize, combines seamlessly the above two dimensions as a writer and an eco-activist. Maathai understanding of the environment is that "it is the place of women farmers with deeply grounded knowledge of ecology and agriculture" (2007, p. 24). In her works *The Green Belt Movement* (2004) and *Replenish the Earth* (2010), Maathai explores the indispensable roles of women in the environment by challenging the subsisting understanding of the roles of women in leadership and environmental consciousness. Maathai's environmental movement organises women in groups with the focus of planting trees and conserving the environment. She championed ecological activism in Kenya by mobilising women to challenge the allocation of public lands to private individuals by the political class.

While she was successful in her resistance against environmental injustice of reducing land to cash value, Maathai rejects masculine ideas of saving the environment because of her conviction that such an option will incapacitate the potential of all humans to protect the environment. Though Maathai (2009, p. 278) argues against the African patriarchal society and the residue of colonialism in land use and allocation matters that favour the men folk, she champions a "genuine partnership that respects gender equity. She advocates that "women can have meaningful lives and feelings of connection and community with each other and with nature and the earth, without depending on men" (Presbey, 2015, p. 44).

Following the above motivation of Maathai, it is important that there is a womanist bias in African environmental ethical discourses. However, this position should not be interpreted as blindly aping the subsisting feminist-ecological trend in Western environmental ethics. The position is informed by the supposition that there might be an important nexus between women and a good environment, especially in the African context. Such supposition is worthy of investigation. Karen Warren (1988), in the Western context, has argued through her eco-feminist thesis that there are strong historical and theoretical connections between the oppressive patriarchal domination of women in Western history and the domination of nature. In the African context, Fainos Mangena (2014) has considered such a possibility in the African context with the conclusion that there is no such connection or relationship, especially in sub-Saharan Africa. He claims that there is the absence of "any connection between the oppression of women of African descent and the ill-treatment of nature" because "nature is owned or guarded by the spirit world" and not possessed or oppressed by the male gender. So for Mangena (2014, p. 1), "ecological feminism is not transferable to sub-Saharan Africa." Without necessarily evaluating the cogency of Mangena's argument, it will be penny wise, pound foolish to consign into silence the salience of the related question of the role of women in solving environmental problems.

Contrary to Mangera's claim of throwing out the baby with the water as a metaphor for women's role and feminist bias, respectively, it is important to consider the connection between women (not feminism) and the environment in Africa. In doing so, understanding the challenges of women is necessary to be able to come to grips with how they can better promote good environmental practices. Molara Ogundipe-Leslie underscores the challenges of African women, which she calls 'six-oppressive conditional mountains' thus:

> One might say that the African woman has six mountains on her back: one is oppression from outside (colonialism and neocolonialism?), the second is from traditional structures, feudal, slave-based, communal etc.; the third is her backwardness (neo-colonialism?); the fourth is man, the fifth is her color, her race; and the sixth is herself.
>
> (Ogundipe-Leslie, 2000, p. 112)

All the above six-oppressive burdens on contemporary women folk in Africa, together with the values in some internal structural issues, must be investigated and understood in future African environmental ethics. While Ogundipe-Leslie has mentioned the external issues, some of the internal concerns deserving philosophical scrutiny include: gendered division of labour, accessibility to and control over material resources such land, water, property, finance and intellectual resources such knowledge, skills and survival strategies, community work and participation in social movements as well as policy matters (Sachs, 1997). These salient issues intersect at different levels.

Beyond the current trend of formulating supposedly unique African environmental ethical theory, scholars working in the discipline should endeavour to connect theory with praxis by studying further the pedagogical impacts of the traditional roles of women as mothers and the role of mothers in environmental education. The images of women as mothers and heritage keepers in traditional and contemporary African societies can be extended to the environment. Conventionally, women as mothers have a deeper insight into the availability, diversity and maintenance of forest resources such as trees, fodders, fuel wood, nuts, fruits, herbs, building materials, medicinal plants and herbs among others (Tiondi, 2000). Indeed, women also do have insights into the domestic functions and risks of water as its carriers, managers and end-users. Leveraging on the significant population of women, as well as on the spill-over effects of the educative potentials of women's environmental consciousness on the younger and future generations, it is a worthwhile engagement to cast a critical searchlight on the values that currently inform (or limit) women in their role as mothers in sub-Saharan African societies.

Conclusion

Future African environmental ethics would profit not only by having a womanist bias for environmental studies but by also appropriating relevant insights that can be further developed from feminist thoughts and practices in Africa. To achieve these related ends, unsystematic and systematic limitations against the strong presence of women's identities in African environmental discourses deserve attention. African and Africanist women scholars working in the field should be encouraged by having their works published in credible academic journals without undermining the quality of research outputs. It is important that a reverse discrimination policy be instituted in recruiting deserving women in academic and research institutions in Africa; this should be done without attenuation for mentoring budding women scholars in the field by established African environmentalist scholars, regardless of gender. The undue concentration on racial identity and the geographical criterion of Africanness of thought should be de-emphasised in African environmental ethics in order to redefine the horizon of the emerging field. A womanist bias is therefore advocated as 'the new turn' in African environmental ethical discourses.

The conversational approach should be sustained in order to further appropriate relevant insights that can be developed from feminists' thoughts and practices in Africa. In this regard, Maathai's views and lived accomplishments are directional. In the light of the salience of binary complementarism as a metaphysical grounding for the discipline of African environmental ethics, males do have tremendous responsibilities and contributions to make to the unfolding womanist struggle for a sustainable environment. For a holistic orientation in African environmental ethics, which will serve as pivot for effective and sustainable management of the environment in 21st-century Africa, the metaphysical must reinforce not only the ethical, womanist relation to the environment, but also non-philosophical variables such as law, technology and education.

References

Agboro, O. P. (2008). African Environmental Ethic: A Creation of Distorted Values. A paper presented at an *International Conference on Rebuilding Africa*, Washington, DC.

Chimakonam, J. (2015). Conversational Philosophy as a New School of Thought in African Philosophy: A Conversation with Bruce Janz on the Concept of 'Philosophical Space'. *Online Journal of World Philosophies*, 3, 9–40.

du Toit, L. (2008). Old Wives' Tales and Philosophical Delusions: On the 'Problem of Women and African Philosophy'. *South African Journal of Philosophy*, 27(4), 413–428.

Ekwealo, C. J. (2011). Environmental Ethics and Values in the 21st Century: An Africanist Philosophical Analysis. *Journal of African Environmental Ethics and Values*, 1, 1–14.

Fadahunsi, A. (2007). *Challenging Africa's Environmental Crisis: The Ethical Imperative*. www.utexas.edu/conferences/africa/2009/09abstracts_a gohtm (accessed July 13, 2012).

Fashola, R. (2008, August 12). Is the Sum Inexhaustible? *The Nation Newspaper*, Tuesday.

Fayemi, A.K. (2013). "A Hermeneutico-reconstructionist Method as a Synthesis of Wiredu's and Oluwole's Methodological Orientations in African Philosophy." A Ph.D. Thesis in the Department of Philosophy, Olabisi Onabanjo University, Ago-Iwoye.

Ikuenobe, P. A. (2014). Traditional African Environmental Ethics and Colonial Legacy. *International Journal of Philosophy and Theology*, 2(4), 1–21.

Kelbessa, W. (2015). African Environmental Ethics, Indigenous Knowledge, and Environmental Challenges. *Environmental Ethics*, 34(4), 387–410.

Maathai, W. (2004). *The Green Belt Movement: Sharing the Approach and the Experience* (Revised ed.). New York: Lantern.

Maathai, W. (2007). *Unbowed: A Memoir*. New York: Anchor/Random House.

Maathai, W. (2009). *The Challenge for Africa*. New York: Anchor/Random House.

Maathai, W. (2010). *Replenishing the Earth: Spirituals Values for Healing Ourselves and the World*. New York: Doubleday.

Mangena, F. (2014). Can Africana Women Truly Embrace Ecological Feminism? Filosofia Theoretica: *Journal of African Philosophy, Culture and Religions*, 3(2), 1–16.

Ogundipe-Leslie, M. (2000). African Women, Culture and Another Development. *Journal of Black Feminism and Social Research*, 27(2), 106–119.

Ogungbemi, S. (1997). An African Perspective on the Environmental Crisis. In L. J. Pojman (Ed.), *Environmental Ethics: Readings in Theory and Application*. Belmont, CA: Wadsworth Publishing Company.

Ojomo, P. (2011). Environmental Ethics: An African Understanding. *Journal of Pan African Studies*, 4(3), 101–115.

Ojomo, P. (2012). Aldo Leopold's Land Ethics: Implications for Africa. A paper delivered at the *International Conference on African Environmental Ethics, Philosophy and Education Versus Environmental Activism and Advocacy: The Great Debate*, University of Lagos, July 11–13.

Oluwole, S. B. (1997). Culture, Gender, and Development Theories in Africa. *Africa Development*, XXII(1), 95–121.

Oluwole, S. B. (2011). *Intellectual Identity and Cultural Diversity*. Unpublished.

Oluwole, S. B. (2012). Sustainable Environment and African Environmental Ethics. A Guest Lecture delivered at the *International Conference on African Environmental Ethics, Philosophy and Education Versus Environmental Activism and Advocacy: The Great Debate*, University of Lagos, July 11–13.

Oluwole, S. B. (2014). *Socrates and Orunmila: Two Patron Saints of Classical Philosophy*. Lagos: Ark Publishers.

Oluwole, S. B., and Sofoluwe, J. A. (2012). *African Myths and Legends of Gender*. Lagos: Ark Publishers.

Osha, S. (2006). Philosophy and Figures of the African Female. *QUEST: An African Journal of Philosophy*, XX (1 & 2), 155–204.

Presbey, G. M. (2015). Women's Empowerment: The Insights of Wangari Maathai. In E. Palmer (Ed.), *Gender Justice: Vulnerability and Empowerment*. London: Routledge.

Sachs, C. (1997). *Women Working in the Environment*. Washington, DC: Taylor and Francis.

Tangwa, G. (2004). Some African Reflections on Biomedical and Environmental Ethics. In K. Wiredu (Ed.), *A Companion to African Philosophy* (pp. 387–395). Oxford: Blackwell Publishers.

Tiondi, E. (2000). *Women, Environment and Development: Sub-Saharan Africa and Latin America*. Thesis, University of South Florida.

Warren, K. (1990). The Power and Promise of Ecological Feminism. *Environmental Ethics,* 12(2), 125–146.

9 Ohanife

An account of the ecosystem based on the African notion of relationship

Jonathan O. Chimakonam

Introduction

I attempt an articulation of a new account of the ecosystem called *ohanife* using the underexplored African notion of 'relationship' found in many places in the sub-Saharan regional cultures and embedded in the Igbo notion of ezi n'ulo, which is emblematic of a relationship of interdependence and complementarity. Here, the ecosystem is conceived as a community and a common home to all there is. It is a network encompassing all existents. In it, every entity, both human and non-human, matters. Everything is connected to this community, called ecosystem, and through it; everything is connected to everything else. In the first part, I will do preliminary conceptualisations. In the second, I will unbundle the theory of ohanife through what can be called the architecture of ezi n'ulo comprising of uloshpere and ilosphere, showing the important place which the notion of relationship occupies in African thought. In the third part, I will schematise what can be called complementary and conversational curves to measure the relationship between oha and ife within the ecosystem. I will then employ these curves to demonstrate that in the ecosystem, everything, whether human or non-human stands equal to others and represents a necessary link in the network of existents and that the rupture of this link might have unknown consequences.

The central point about the philosophical discussion of the ecosystems is 'relationship.' The answer to what the nature of the relationship between the components of the ecosystem should be is cardinal to understanding the ethical issues associated with the ecosystem debate. Why should I as a human being treat other non-human entities in the ecosystem with respect or pay some duty to them as against my economic interest? This is the big philosophical question which environmental ethicists are confronted with in the discussion about the stability of the ecosystem. The anthropocentric view places humans at the centre of the ecosystem and treats other entities in it as serving the needs of humans. They are not ends in themselves; they are means. As a result, humans do not owe them moral obligations as such (Kant, 1994, p. 28; Warren, 1994, pp. 49–51).

Debates on whether humans are in any way obligated to non-human members of the ecosystem have given rise to many concepts, such as sentience, reason, interest and inherent value. Almost all of these have been vitiated one way or the other. The fate of the non-human members of our common earth continues to hang in the balance. But one must ask: why should I protect the environment and other non-human members of the ecosystem? Some argue that we must protect the environment because we owe a moral obligation to posterity, a duty to leave the earth habitable and capable of supporting life for the benefit of posterity. This truly sounds nice, but the question a critic can quickly raise is: what has posterity ever done for me to deserve such a duty from me? (The above is not a stupid question. It is perfectly reasonable and I should add that any human has the right to ask such a question and demand an answer.) But the truth is that posterity has never done anything for anyone to deserve such an obligation from them. However, this may not be the only question. Consider an alternative: what can I do for posterity? This equally makes great philosophical sense because in it, we are involved. Posterity does not yet exist. It has neither will nor power, but we do have all those things. We can meaningfully talk of what we can do for posterity, whereas they cannot talk of what they can do for us. We cannot even meaningfully talk of what they can do for us. So, one thing which we can do is to ask what we can do for posterity. It is in this connection that the idea of protecting other members of the ecosystem as a duty to posterity rings loud. But should this be the right motive?

The position of this work is that all motives that are anthropocentric cannot qualify as the right motive for the argument in favour of environmental protection or the conservation of the non-human members of the ecosystem. Basic in the discussion involving human and non-human members of the ecosystem is relationship. It is what is considered to be the structure of the relationships between humans and non-humans that characterise different conclusions scholars reach on the subject of ecosystem and its conservation. Those who think that this relationship is an 'end-means' affair treat humans as subjects that are ends in themselves and treat non-humans as objects that are meant to serve as a means for reaching human ends. It is positions such as this that inspire anthropocentric visions in environmental ethics. Contrary to the preceding, I am inclined to think of this relationship as an 'end-end' affair. Different scholars have held this position in one way or the other. For example, Bernard Rollin founded this equality on the notion of sentience, which, he says, is shared by both humans and non-humans (1994, pp. 29–32). Peter Singer in talking of animal liberation vigorously campaigned for equal treatment for humans and animals and grounded this equality on the ethical principle of utilitarianism. He eminently criticised what he calls "the tyranny of humans over non-humans" (1975, p. 17). Tom Regan appears to reject Singer's utilitarianism for lacking the notion of intrinsic worth and calls for radical egalitarianism in which there will be total respect for non-humans on the basis of their inherent value (1994, pp. 40–46).

The notion of inherent value is subsumed under the so-called sentience-based ethics, which contrasts with anthropocentric ethics in the debate on the ecosystem. Those who argue for the equality of all components of the ecosystem usually follow the path of sentience-based ethics. Those who are opposed to this adopt anthropocentric ethics. All of these people are united in their thinking that some form of equality characterises this relationship.

What is different, however, between this mainstream left-wing movement (those who support an egalitarian approach to the environmental protection debate) and my position is that they may be described as soft-leftists (those who marginally support the egalitarian position in the environmental protection debate) because they tend to demand an appreciable level of respect for non-humans that will impel us to protect and conserve them but not necessarily a recognition of any form of ontological equality between humans and non-humans. From the recognition of their ontological variance, they assume their ontological inequality, but without stating it. It is this mindset that undergirds their guided and cautious soft-leftist stance. On the other hand, I recognise their ontological variance but also recognise their ontological equality. The fact that trees are different from humans in their ontological structure does not mean that the property of treeness is inferior to the property of humanness in the ecosystem. I say that they may be different, but they stand as equals in the ecosystem and this position may be described as 'hard-left' (those who strongly oppose the non-egalitarian stance in the environmental conservation debate). Generally, leftists in the left-right political spectrum are movements of change opposed to order and convention. Order and convention represent the right-wing.

Also, the soft-leftists, as I describe them above, appear to supply no clear tool for measuring this relationship of equality. It remains a mere suggestion that can easily be readjusted as stakes grow higher. On the contrary, I formulate the tools of complementary and conversational curves. By the former, I will characterise and demonstrate this relationship between two equals as complementary. And by the latter, I will characterise and demonstrate the process of this relationship as conversational, i.e., something that is balanced and dispassioinate. It is important to explain that a conversational relationship is not the same as Hegel's dialectic relationship, which creates a synthesis from thesis and anti-thesis. A conversational relationship does not create a synthesis; it reshuffles a new thesis and anti-thesis from a given thesis and anti-thesis. In other words, a conversational relationship is somewhat dialectic, but a dialectic relationship is not conversational. The conversational relationship which I formulate here is driven by two types of motions, namely disjunctive and conjunctive. The former sets opposed variables apart, while the latter brings them together. From a disjunctive motion, a dialectic concession emerges to transform the motion into a conjunctive one where, again, a dialectic complementarity emerges to transform the motion yet again to a disjunctive one. This is a shuffling of thesis and anti-thesis which sustains the conversational relationship.

In this work, I will first unpack the concepts that will drive my thesis of *ohanife*. I will also unfold the thesis of *ohanife* through the under explored African notion of relationship embedded in the Igbo idea of *ezi n'ulo* architecture. I will schematise the complementary and conversational curves as viable tools for measuring relationships and the process of these relationships between two seemingly opposed variables. I will then deploy these tools in assessing the relationship between humans and non-humans within the ecosystem. I will then conclude my proposal of ohanife as an account of the ecosystem by submitting that this relationship is complementary and that its process is conversational.

Ohanife: unpacking the main concepts

In this work, I shall be introducing a number of concepts into the field of environmental studies. This is necessary for the project I wish to undertake, i.e., articulating a theoretic account of the ecosystem from an African perspective. Some of these concepts include ohanife, the architecture of ezi n'ulo, ilo, ilosphere and ulosphere.

Ohanife: This derives from two Igbo words, namely Oha, which means people usually in a group, community or network, and ife or ihe, which means things, usually including both living things that are not humans and non-living things. A broad term for ife would be non-humans. When put together, ohanife simply means a network or community of humans and non-humans, in other words, ohanife as a new coinage becomes the Igbo synonym for the ecosystem, and this is the meaning I shall attach to it in this work.

Ezi n'ulo architecture: this concept derives from three words, two from the Igbo language and one from the English language. Ezi n'ulo is formed from a conjunction of two Igbo words ezi and ulo. Ezi means space or environment, usually for non-humans. Ulo also means space or environment for humans. The English word architecture etymologically comes from the Latin *architectus*, which means the science of designing and building structures (Smith, 2004, p. 76). Put together, ezi n'ulo architecture simply refers to the structure of ezi n'ulo. And in this context, that would be the architectonics of the interactive environments of humans and non-humans, i.e., ohanife or the ecosystem.

Ilo: Let me begin by saying that ilo in the Igbo language is a synonym for ezi. Incidentally, it is a synonym that is generally preferred among Igbo thinkers in discourse theories to its archetype, which is ezi. So, ilo has a more popular usage than ezi. In this work, I am going to employ ilo more often than ezi. Like ezi, it means a space or an environment for non-humans. However, in recent usage, especially associated with the internet and social media, a number of Igbo thinkers are rethinking the concept in terms of virtual space or environment for human interactions (Oha, 2016, p. 10). What is important is that these avant-garde thinkers retain the notion of interaction, which is cardinal to the ilo environment, only this time, they substitute non-humans for humans in what one may describe as the human invasion of the ilo. But there is something unorthodox here. The virtual ilo is not exactly like its archetype, which is a physical space of physical objects, even though it simulates the physical ilo. In the virtual ilo, members are subjects, whereas in the physical ilo, they are objects. But because the virtual ilo does not have the same architectonics as the physical ilo, the subjects in the virtual ilo become objectified. This shows that ilo, no matter the type of restructuring it undergoes, is likely going to remain an environment for interactive objects. The metaphysical implication is that environment determines ontology. Elements in what can be called the ilosphere are objects because the ilo environment determines them to be so. Similarly, elements in the ulosphere are subjects because the ulo environment structures them to be so. It is not consistent with the logic of ezi n'ulo architecture for members of one environment to become members of the other and vice versa, even though they can interact and complement each other. Thus rethinking the architecture of ilo in terms of the cyberspace, even though it licenses humans to interact in such an environment as members, the bigger picture shows that it also reduces them to objects. We do not perceive humans in the cyberspace; we perceive only their activities and activities are not subjects. I needed to make this clarification before I go on with the unbundling of the ezi n'ulo architecture in light of environmental thought.

Ilosphere: This concept derives from two words from two different languages, namely Igbo and English. We have said enough about the Igbo word ilo already. The second root word of the concept of ilosphere is the English word sphere. It comes from the Greek word *spharia*, which means globe or ball. A sphere can be defined as "a geometrical figure that is perfectly round, 3-dimensional and circular – like a ball" (n.a, 2016, http://study.com/academy/lesson/sphere-definition-formulas-quiz.html). From the foregoing, I conceive of an ilosphere geometrically as the set of all points equidistant from a single point in the ezi space that has non-humans as its elements and where the ezi n'ulo architecture represents a ball encapsulating the ilosphere. Mathematically, a sphere is different from a ball in that, whereas a sphere can be seen as a two-dimensional closed surface embedded in three-dimensional Euclidean space, a ball is a three-dimensional shape that includes the sphere together with every other thing within the sphere (Smith and Vamanamurthy, 1989). In this way does the ezi n'ulo architecture include the ilosphere and all of its elements.

Ulosphere: This concept is also derived from two words from two different languages. The Igbo word ulo means house, home or simply the human environment. The English word sphere which comes from the Greek root word spharia means, as we have stated earlier, globe or ball. Needless to repeat the definition of sphere, I conceive of the ulosphere geometrically as the set of all points equidistant from a single point in the ulo space that has humans as its elements and where the ezi n'ulo architecture represents a ball encapsulating the ulosphere. This implies that the ulosphere is within the ezi n'ulo architecture, just like the ilosphere. In the next section, I will attempt to unpack ohanife through the ezi n'ulo architecture.

Understanding ohanife through the ezi n'ulo architecture

My attempt in this work is to give an account of the ecosystem using an underexplored African notion of relationship. The notion of relationship or communion is central to African thought in that it characterises life in many cultures in sub-Saharan Africa (Metz and Gaie, 2010, Metz, 2011, 2013, 2015, 2017). The Igbo notion of ezi n'ulo is one of the structures which can be used to demonstrate the practical effect of relationship. It specifically harps on the idea of complementarity and mutual dependence. In traditional Igbo architecture, a comprehensive design of an abode necessarily includes spaces for humans, non-humans and even the spirits alike. This suggests an admission of their interrelatedness and interdependence. Hence, the concept ezi n'ulo captures these two abodes. I will seek to unravel this structural unity from the perspective of interdependence between oha (humans) and ife (non-humans). Ohanife therefore becomes the Igbo word for the ecosystem. One way to understand the structure of the ecosystem from an African point of view would be to understand the architecture of ezi n'ulo. It has three main components of the traditional Igbo architectural design, namely, an abode for humans, the abode of other living and non-living things occupying the space outside the human abode and, finally, the abode of the family guardian deity, which is located in front.

If we describe the abode of the humans as the ulosphere, i.e., the spaces and their appurtenances in which humans reside and carry out their earthly activities, we will describe the abode of the non-humans as the ilosphere and that of the family guardian deities as the agwusphere. We must note at this point that when we talk of 'sphere' in this context, we mean, broadly, something close to all 'spaces of influence' within the ecosystem and not merely a living place. Thus taken under this interpretation, one can easily see the interconnection among these three components

Thus there is a mutual dependence among the three. While the two innermost abodes in the design depend on the abode of the guardian deities for their protection, the abode of the guardian deities depends on them for its nourishment. Also, while the human abode depends on the non-human abode for nourishment and sustainability, the latter depends on the former for conservation and nourishment. So, there is a mutual interdependence among these three components.

However, I am developing an empirical theory of the ecosystem with a universal appeal. This means that I am going to chop off the spiritual component of the Igbo architecture in order to deal clearly with the relationships among physical realities that make up the ecosystem. The above structure will then look like the diagram below in the new system I aim to formulate.

From the above, it is easy to see how the traditional Igbo architecture transforms to the structure of the ecosystem or ohanife in which there is mutual interdependence among all physical realities, both humans and non-humans alike. In the next section, I will attempt to present clearly, the sort of relationship that underpins the interaction among all realities in ohanife. I describe this relationship as complementary.

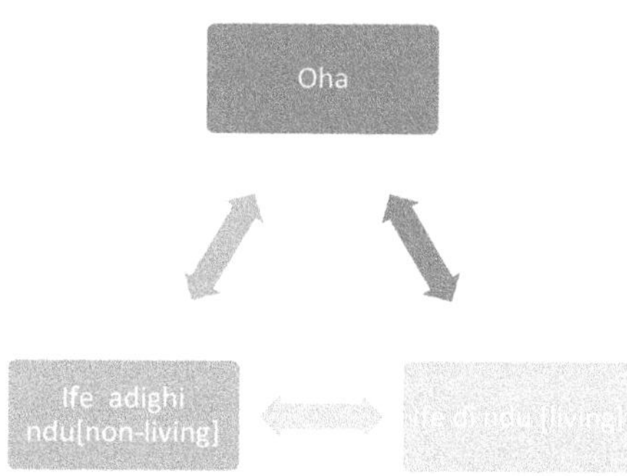

Figure 9.1 A relational structure of the ecosystem

The complementarity of oha and Ife

The ecosystem is seen as a network of all there is – a community of some sort (Leopold, 1994, pp. 84–91). The fuss, however, is that anthropocentrists believe that humans – a component of the community – are not only superior but should be dominant (Watson, 1994, pp. 117–123). The account we are building here contradicts the anthropocentrists. It holds that realities in ohanife fall under two broad categories, namely, oha or humans and ife or non-humans. Ife could further be divided into two, i.e., ife di ndu or living things and ife adighi ndu or non-living things. Of these divides, ohanife holds that there is no superior/inferior or dominant/dominated dichotomy in their relationship with each other. Everything in the ecosystem serves a necessary link of reality. This is the statement of the principle of nmekoka or complementarity (Chimakonam, 2014). This can be contrasted with Innocent Asouzu's idea of complementarity, formulated in terms of the missing link of reality (2004, p. 320). That something is a necessary link implies that the presence of it creates completeness, whereas that something is a missing link implies that the absence of it creates incompleteness in the chain of reality. Both may suggest the idea of relativity in ontology, but while the former suggests the relativity of presence, the latter suggests the relativity of absence. But scholars of the conversational mindset believe that it is better to talk of reality as a presence rather than as an absence, hence, the idea of the necessary link (Chimakonam, 2013, p. 79).

Oha and ife inevitably complement each other. The coinage suggests this. Neither oha nor ife can thrive without the other. The question then is: how do they complement each other? One way to read this is to imagine the patterns of ulosphere and ilosphere as two circles, the former smaller and embedded within the latter, which is bigger. One could say that, while ulosphere is at the centre, ilosphere is at the periphery. Two ideas are suggested, namely: (a) that the central sphere appears dominant and (b) that the peripheral sphere appears to dominate the centre. Another, perhaps more complementary, way of looking at it is to see their mutual interdependence. One may then interpret ulosphere as the 'centre' of the periphery and ilosphere as the 'periphery' of the centre. Thus, while the centre conserves the periphery, the periphery sustains the centre.

Ohanife projects the idea that the ecosystem is a community and therefore the common home of all there is. The reason for this conception is embedded in the notion of relationship in African thought, which presents things in the world as interconnected in a network-like system. Every entity in the system, whether human or non-human, matters. Everything is connected to this community called ecosystem and through it; everything is connected to everything else. This is the main thesis of ohanife articulated here as an account of the ecosystem.

We can think of the African understanding of community in this regard. A territorialised environment consisting of ife in which there are abodes occupied by oha with a common socio-political and economic destiny and maintained by a body of norms. In this way, the community is not just a space for humans, and humans are not the only residents of the community. Unfortunately, some African scholars, in theorising about the community, have spoken of humans as the only entities in the community. This is because; they look at the world only from anthropocentric perspective. Some of the theorists guilty of this lopsided view of the community include Menkiti (1984, pp. 171–180), Gyekye (1997, pp. 101–122), Ikuenobe (2006, pp. 117–131) and Matolino (2014, pp. 160–186). It is a thing of surprise that these scholars suggest that oha can truly survive in the community without ife. Ife is necessarily part of the community as much as oha is. Whatever oha does in the community, one way or the other, there is a path which it crosses with ife. In fact, ife is strategic to the design and activities of oha. It is the sustainer of oha. But as ife sustains oha in the community, it falls on oha to conserve ife. So, there is a duty-like relationship. What, however, makes that duty appears lopsided is that the architecture of ilosphere is not a duty-demanding agency. Trees, for example, will not ask us not to fell them by arguing that they have kept their duty to us, nor will a coral reef protest against a planned dredging. The reason is that ife is not in the same linguistic order with oha. In other words, the language of the ilosphere is not the same as the language of the ulosphere. But there is a language common to both, i.e., sign language, and ife utilises this to communicate to us, whether in the chirping of the birds, the gestures of the peacock, the whistling of the pine or the swinging of the palm frond. Incidentally, oha has been poor in developing this communication mechanism with ife since the dawn of evolution.

Again, the architecture of ife is designed to discharge its duty to oha without prompting, so oha assumes quite in error that ife, possessing no will, deserves no duty in return. The rain will fall in the rainy season and the sun will shine in the dry season just as the fruit trees will yield in their respective seasons. Animals will reproduce their kind in abundance for the nourishment of oha and the river will be there the next morning for oha to drink from. There is indeed no attitude from ife that will compel oha to pay some respect. But this is where the problem lies. Duty is not something paid because there is a presentment of repercussion or some utility. Unfortunately, that is how some Western environmentalists like Martin Golding (1994, pp. 225–230) understand it. In the account of the ecosystem I promote, duty is a necessary and complementary debt which every entity owes to the other. This is in recognition that no single entity in the ecosystem is self-sustaining. Each entity's ability sustains or conserves another and the motto of ohanife therefore is: 'let every entity do what it is naturally designed to do and all will be well.' In other words, oha need not abandon its duty to ife simply because ife does not demand it vocally or by means of an obvious threat. If oha claimed that it possessed the faculty of reason which none other in the ecosystem possessed, then let it utilise that faculty to recognise the necessity of mutual complementation of duty. This fact appears to be recognised by other entities in the ecosystem which do not even possess the faculty of reason. The formation of ife architecture clearly obeys the principle of nmekoka. Thus our agitation for the stability of the ecosystem should ultimately be a counsel for oha to obey the principle of nmekoka. When things fall apart in the architecture of the environment, the only entity to be blamed is oha. It does appear that this is why it alone possessed the faculty of reason. It alone truly owes a duty in a normative sense to ife. Thus, one question that looms concerning the human relation to the environment is: 'how well have we utilised our faculty of reason'?

Attempting to answer the question above is beyond the scope of this work. The point I would rather wish to make is that there is a complementary duty between oha and ife but, *a fortiori*, that duty is a necessary one. While oha owes ife a normative duty as an entity possessing the faculty of reason, ife which does not possess this faculty owes oha what can be called 'design-duty.' What this means is that ife is designed by nature to follow a certain pattern and to serve the need of oha. Now, whenever and wherever ife fails in this design, it loses its value. We were told in the Bible story that a certain Jesus of Nazareth cursed a fig tree for failing to produce fruits at the right season that nature has designed it to do so. Thus, for failing in its necessary design-duty to oha, it had to be cursed.

To properly describe the necessity which undergirds the complementary relationship between oha and ife, I will employ the notion of complementary curve. The idea of using curves to measure relationships, which is commonplace in economics, has not been a feature in philosophy. It is interesting to see that it is not a bad idea for philosophy after all. I will here give the concept proper definition, articulation as a tool for measuring the relationship between seemingly opposed variables, and graphic representation. This will transform it into a powerful tool that can always be appealed to in explaining any relationship involving two opposite entities or ideas.

Definition: The complementary curve is a graphic representation of the necessity of relationship between one variable and its opposite. It is drawn with the motion of the variables on the vertical axis and the variables themselves on the horizontal axis. The complementary curve has six geometric components representing variables, motions and relationships. They are: two right angles representing two opposite variables we may symbolise as 'S' (self) and 'O' (other); two motion indicators called the disjunctive motion represented by logical connective for disjunction (V) and the conjunctive motion represented by the dotted connective for conjunction (Λ); the dotted horizontal line on top of the dotted conjunction symbol representing the complementary bar; and, finally, the dotted face-down K sign representing Conversational Dialectics (a point that characterises the encounter/interaction between two opposed variables) within which Dialectic Complementarity and Dialectic Concession occur only to be unsettled yet again by Tension of Incommensurables that sustains and drives the conversation. Below are diagrams of the complementary curve and conversational curve.

However, the complementary curve is subsumed within the conversational curve. This is because proponents and opponents as we have it in conversational philosophy must first converse before they can complement. In other words, without conversation, there cannot be complementation. Thus, the conversational curve encompasses the complementary curve or, put differently, the complementary curve is a component of the conversational curve. I will immediately define and diagram the conversational curve.

Definition: The conversational curve is a graphic representation of the dialectic relationship between nwa-nsa (proponent of a view or thought) and its nemesis

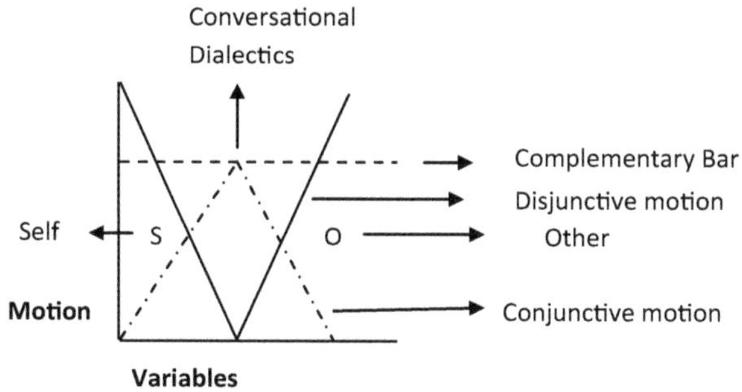

Figure 9.2 Diagram of complementary curve

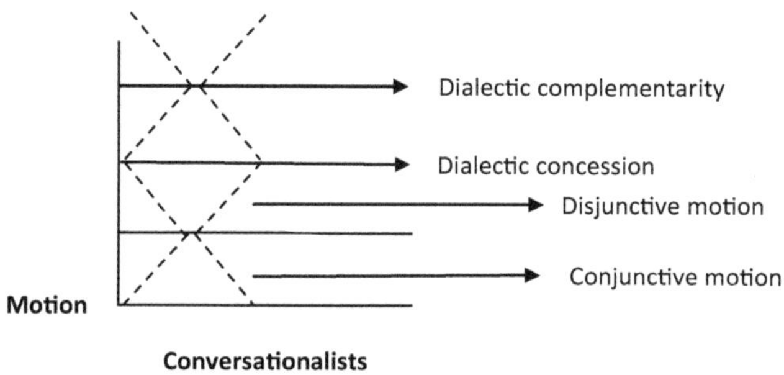

Figure 9.3 Diagram of conversational curve

nwa-nju (opponent of a view or thought). It is drawn with the motion of conversation on the vertical axis and the conversationalists themselves on the horizontal axis.

A look at the conversational curve will immediately reveal the dotted disjunctive and conjunctive motions of the seemingly opposed variables. For the oha as well as some components of ife that are capable of movement, this motion is a geometric one, i.e., from point to point. But for the other components of ife that are not capable of movement, this motion is like that of the sun. We say, for example, that the sun rises in the East and sets in the West, but, in actual astronomical terms, we know that the sun is fixed in its galactic position. We know that the planets are the ones that revolve and orbit around it. For Earth in particular, we know that its movement around the static sun is critical to the sustenance of life on it. Similarly, the ife architecture is critical to the sustenance of the oha in the ecosystem.

The conversational relationship is different from the dialectic relationship, even though there is a marginal bond between them. One can say that the conversational relationship is dialectic since it accommodates the conflict between thesis and anti-thesis but it cannot be said that the dialectic relationship is conversational. The latter impossibility is due to the fact that while dialectic relationship yields a synthesis, the conversational relationship reproduces another round of thesis and anti-thesis at a higher level. It is my thinking that the dialectic relationship is not the ideal philosophical relationship. Philosophical answers are not supposed to come in the mode of syntheses. In fact, it would take an overdose of compromise to create a synthesis. And synthesis is very much like consensus. George Hegel got it wrong after all. We know him to be associated with quite intrusive errors in philosophy. His philosophy of history, for example, is littered with great errors. Karl Marx, who copied the Hegelian dialectics in formulating his theory of economic history, also committed his theory to the problem of the dialectics.

Philosophical answers are not supposed to be syntheses (consensus) that morph into another thesis; this is not so coherent. First, dialecticians did not tell us how synthesis could transform into a new thesis. Second, they could not tell us how philosophical consensus, which synthesis at least approximates, could become possible. Third, even if philosophical consensus becomes possible, should it not be unassailable? If all philosophers agree, who would then disagree? Put differently, if two opposed variables arrive at a synthesis, how can this, in practice, possibly become a new thesis? This is why synthesis, which is somewhat consensual, should be an enemy of philosophy. Marx's economic ideology is one example where Hegel's dialectics have been put to practice. What is more, the dialectics which yield a synthesis that gets transformed into a new thesis are yet to come through successfully.

It is on this ground that I have formulated the conversational relationship to make up for these loopholes. In it, synthesis is eliminated. Philosophical answers come in the form of new theses that keep being reformulated at new levels after each creative struggle between thesis and anti-thesis. The disjunctive and conjunctive motions that drive this process never really go too far apart and never really come too close together, respectively. This is however not the case in a dialectic relationship in which the supposed conflict between thesis and anti-thesis crystallises into both, completely intersecting to create a new order. The diagram below is for the dialectic curve.

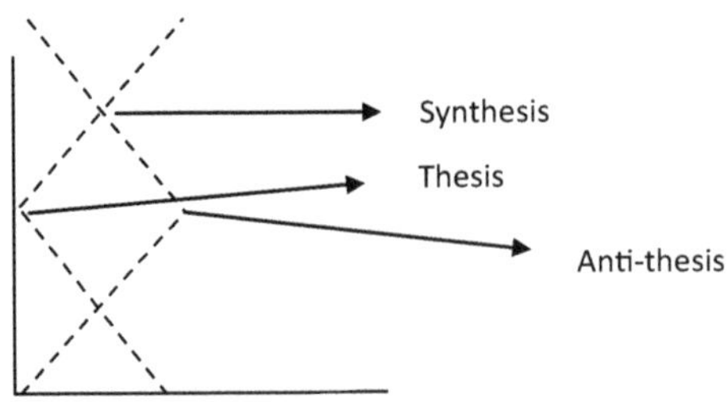

Figure 9.4 Diagram of dialectic curve

The dialectic curve by definition is supposed to be the graphic representation of the relationship between two seemingly opposed variables. The problem, however, is that it yields false results but portrays them as if they were true. In the above, we can see that the motions of thesis and anti-thesis criss-cross to yield a synthesis and then continue on to reproduce another round of thesis and anti-thesis from the synthesis. How can two opposed variables completely agree? It is remarkably unbelievable that, for ages, philosophers have picked very little quarrel with this false structure. Anyone still itching for a proof that the dialectic relationship is philosophically incorrect should look at the lame Hegelian absolute spirit or the elusive prophesies of Marxism. My discussion of this will be the focus of another work.

The conversational curve as a correct representation of the philosophical relationship thus demonstrates that every variable within a system of opposition constructs a pattern of motions for itself in relation to the other. When entities in the ecosystem are subjected to this conversational measurement, they manifest both disjunctive and conjunctive motions. A disjunctive motion reveals difference (a form of ontological variance), while a conjunctive motion reveals identity (a form of ontological equality). A disjunctive motion culminates in what is called dialectic concession in which seemingly opposed variables recognise the place, importance and contribution of one another (Chimakonam, 2015, pp. 470–471). A conjunctive motion, on the other hand, culminates in what is called dialectic complementarity in which seemingly opposed entities mutually complement one another in the mode of from each according to its best abilities (Chimakonam, 2015, pp. 470–471).

One can also observe from the conversational curve above that, as soon as opposed variables arrive at the dialectic complementarity, they soon break off into a disjunctive motion once again. This is caused by what is called the tension of incommensurables (Chimakonam, 2015, p. 468). In the analysis of the ecosystem, oha and ife have a different ontological architecture, for example, treeness is not the same as humanness (ontological variance) even though treeness is not inferior to humanness (ontological equality) in the ecosystem. Thus, despite the necessity for their interdependence, their interaction is always intermittent due to their ontological variance. For example, a poultry farmer may have the need to look after her birds and feed them, but she could not choose to eat the mash and sleep in the same place with them. Thus while shared identity brings opposed entities together, it is their difference that always breeds tension and intermittently sets them apart. This thing called tension of incommensurables is what transforms dialectic complementarity into a disjunctive motion in the relationship of seemingly opposed variables.

From the foregoing, one can observe that, in the ecosystem, oha and ife components, despite having a different ontological architecture, inevitably have cause to complement one another from time to time. My position is that this complementation is a necessary one. And this is because neither oha nor ife, which occupy the ulosphere and the ilosphere respectively, can ultimately be without the other.

These two spheres are inseparable. The interaction of their occupants is thus a dialectic one that rides on the crest of disjunctive and conjunctive motions. The relationship between oha and ife, as shown in the curves, is mutual, balanced and equitable. It is not the one of a superior-inferior dichotomy. It is marked by mutual interdependence, interrelation and complementation. The idea held by some philosophers in the broad field of environmental ethics, which projects an anthropocentric view of the ecosystem, is not only false but misguided. Any such view is a product of ignorance of the architecture of the ecosystem. The account which I have tried to weave here deriving inspiration from the African notion of relationship, which the Igbo concept of ezi n'ulo replicates, is aimed at exposing the inherent ignorance in the anthropocentric view of the universe as a whole.

Conclusion

In this work, I have derived inspiration from an underexplored African notion of relationship to develop a new relational account of the ecosystem. In it, I have divided the entities in the ecosystem into two broad components, namely oha, or the humans, and ife, or the non-humans. I have further divided ife into two parts, that is, ife adighi ndu, or the non-living things, and ife di ndu, or the living things. I have shown that the difference in the ontological structure of oha and ife does not suggest a superior-inferior dichotomy but merely difference in architectural spheres. I have called these spheres ulosphere and ilosphere and demonstrated how the relationship of oha and ife, which can be measured with a conversational curve in terms of dialectic concession, dialectic complementation and tension of incommensurables, translates into a form of correlation between ulosphere and ilosphere. I have employed a complementary curve and a conversational curve to demonstrate that the relationship and the process of this relationship between oha and ife, as, indeed, all seemingly opposed variables that have this same structure, is that of mutual and balanced complementation rather than a superior-inferior divide, as the anthropocentric philosophers in environmental ethics would suggest. What these curves show us in the final analysis is that, in the ecosystem, everything, whether human or non-human, stands equal to others and represents a necessary link in the network of existents and that the rupture of this link might have unknown consequences.

References

Asouzu, I. I. (2004). *The Method and Principles of Complementary Reflection in and Beyond African Philosophy*. Calabar: University of Calabar Press.

Chimakonam, J. O. (2013). Integrative Humanism: Extensions and Clarifications. *Journal of Integrative Humanism*, 3(1), 73–83.

Chimakonam, J. O. (2014). Ezumezu: A Variant of Three-Valued Logic. Paper presented at the *Philosophical Society of the Southern Africa PSSA*, Free State University, Bloemfontein, South Africa, January 20–22.

Chimakonam, J. O. (2015). Transforming the African Philosophical Place Through Conversations: An Inquiry Into the Global Expansion of Thought (GET). *South African Journal of Philosophy*, 34(4), 462–479. http://dx.doi.org/10.1080/02580136.2015.1104795

Golding, M. (1994). Obligations to Future Generations. In L. Pojman (Ed.), *Environmental Ethics: Readings in Theory and Application* (pp. 225–229). Boston, MA: Jones and Bartlett Publishers.

Gyekye, K. (1997). *Tradition and Modernity: Philosophical Reflections on the African Experience*. Oxford: Oxford University Press.

Ikuenobe, P. (2006). *Philosophical Perspectives on Communalism and Morality in African Traditions*. Lanham: Lexington Books.

Kant, I. (1994). We Have Only Indirect Duties to Animals. In L. Pojman (Ed.), *Environmental Ethics: Readings in Theory and Application* (pp. 27–28). Boston, MA: Jones and Bartlett Publishers.

Leopold, A. (1994). The Land Ethic. In L. Pojman (Ed.), *Environmental Ethics: Readings in Theory and Application* (pp. 84–91). Boston, MA: Jones and Bartlett Publishers.

Matolino, B. (2014). *Personhood in African Philosophy*. Pietermaritzburg: Cluster Publications.

Menkiti, I. (1984). Person and Community in African Traditional Thought. In R. A. Wright (Ed.), *African Philosophy: An Introduction* (3rd ed., pp. 171–181). Lanham, MD: University Press of America.

Metz, T. (2011). Ubuntu as a Moral Theory and Human Rights in South Africa. *African Human Rights Law Journal*, 11, 532–559.

Metz, T. (2013). The Western Ethic of Care or an Afro-Communitarian Ethic? Specifying the Right Relational Morality. *Journal of Global Ethics*, 9(1), 77–92. http://dx.doi.org/10.1080/17449626.2012.756421

Metz, T. (2015). An African Egalitarianism: Bringing Community to Bear on Equality. In G. Hall (Ed.), *The Equal Society: Essays on Equality in Theory and Practice* (pp. 185–208). Lanham, MD: Lexington Books.

Metz, T. (2017). Replacing Development: An Afro-Communal Approach to Global Justice. *Philosophical Papers*, 46(1), 111–137. http://dx.doi.org/10.1080/05568641.2017.1295627

Metz, T., and Gaie, J. B. R. (2010). The African Ethic of Ubuntu/Botho: Implications for Research on Morality. *Journal of Moral Education*, 39(3), 273–290. doi: 1080/03057240.2010.497609

N.A. (2016). *Sphere*. http://study.com/academy/lesson/sphere-definition-formulas-quiz.html (accessed April 23).

Oha, O. (2016). Communicating Democratic Values in a Changing African World. Paper presented at the *Faculty of Arts International Conference*, University of Calabar, April 12–15.

Regan, T. (1994). The Radical Egalitarian Case for Animal Rights. In L. Pojman (Ed.), *Environmental Ethics: Readings in Theory and Application* (pp. 40–46). Boston, MA: Jones and Bartlett Publishers.

Rollin, B. (1994). Sentience Is the Criterion for Moral Worth. In L. Pojman (Ed.), *Environmental Ethics: Readings in Theory and Application* (pp. 29–32). Boston, MA: Jones and Bartlett Publishers.

Singer, P. (1975). *Animal Liberation*. New York: Random House.

Smith, D. J., and Vamanamurthy, M. K. (1989). How Small Is a Unit Ball? *Mathematics Magazine*, 62(2), 101–107.

Smith, S. (Ed.) (2004). *The New International Webster's Comprehensive Dictionary of the English Language: Encyclopedic Edition*. FL: Trident Press.

Warren, M. A. (1994). Difficulties With the Strong Animal Rights Position. In L. Pojman (Ed.), *Environmental Ethics: Readings in Theory and Application* (pp. 47–52). Boston, MA: Jones and Bartlett Publishers.

Watson, R. (1994). A Critique of Anti-Anthropocentric Biocentricism. In L. Pojman (Ed.), *Environmental Ethics: Readings in Theory and Application* (pp. 117–123). Boston, MA: Jones and Bartlett Publishers.

10 Hermeneutics of trees in an African context

Enriching the understanding of the environment 'for the common heritage of humankind'[1]

Angela Roothaan

Introduction

In this chapter, I will look into case studies of the relations of human communities to trees, especially in West-African traditions, and ask how these contribute to a richer understanding of the environment. In traditional cultures all over the world, trees have been important to humans as 'natural symbols' of the central values of communal life, as sources of food and medicine, and as signs of spiritual realities. African traditional relations to trees are of special interest as they are contested in present-day clashes between modernisation movements, religious conversion movements, secular conservationalist movements, and traditionalist movements. Looking into the content of the contested – a rich hermeneutics of trees – is not only of historical interest. It can also contribute to an understanding of the environment that can help create more sustainable relations of humans to their environment for the future.

In the first section, I will outline which different frameworks determine the human relationship to the environment, especially in the West-African context. The frameworks described are those of secularisation, of monotheistic theology, and of traditionalism. Whilst conservationalists might work together with Christians, Muslims, or those practicing traditional religions, their basic presuppositions will be scientific and secular. All the same, Muslims and Christians have their own spiritual understanding of the relation of humans to nature, in the frame of a theology of creation, religious morality and the afterlife, which can conflict with secular as well as traditionalist outlooks. All parties mentioned will also try to negotiate their interests within the framework of the global economy, 'selling' sacred forests as places of touristic interest or of rare plant life, for instance, or using religious fervor to promote economic progress.

After presenting the intersections of frameworks for understanding the meaning of trees, I will discuss, in the second section, two examples of cases of the cutting of trees which show conflicting meanings attributed to trees. Here I will show how the frameworks clash into each other and make a shared discourse on what is of common human interest a complex issue. To view trees as a place of contested meaning, it is important to understand them as bearers of symbolism and signification – something which has been described in the work on trees in public and religious discourse in Senegal by geographer and Islamologist Eric Ross. In the third section, I will go deeper into his work on the holy city of Touba, the capital of Mouride Sufism, which represents tûbâ, the tree of paradise. This work shows us how neoplatonic understandings of (spiritual) reality play their role in Sufi theology, and how an ontology can be seen to be at work in it (as it is in Christian traditions that value worship of holy places, moments, and persons), which I will call 'shamanistic.'[2] Ross has thus shown that religious ascription of meaning to trees is not to be understood in a simple, one-dimensional manner, but that, here, as in other frameworks, complex historico-cultural developments have shaped understandings and practices.

In the final section, I will then return to a more general discussion of how different contexts and discourses lead to contested meanings of trees. I will present here also a more specific philosophical account of African views of nature, as it is given in an article by Michael Onyebuchi Eze on eco-humanism. In the tradition of writers such as Tempels and Mbiti, he proposes to adopt a holistic view of nature understood as life force, which leads to viewing human beings as an element in the whole of nature. As much as such an account helps to explain the spiritual meaning of trees, it suppresses the difficulties described in my earlier sections, arising from the clashing of frameworks. Therefore, I make a plea for a multidimensional approach, which brings the varying positions and views into dialogue after first having disentangled and understood their differences and conflicts.

Intersecting frameworks

When we study the subject of trees in an African context, several frameworks of action have to be taken into account. There is, first, the framework of modernisation, with its separate streams of secularisation, the growth of an autonomous economic sphere and technological development. These streams are often taken to be identical, but although they have been going hand in hand for the most part of European history since the 17th century, they never have been simple, and their relations are certainly more complex in other regions and times. In the United States, for instance, for a long time one of the world's driving nations with respect to economic and technological developments, society and politics remained largely determined by the Christian religion.

On the African continent, things have been still more complex, and historiographies which are not biased by the colonial outlook are still in the process of being written. Although secularisation is a force in present developments, the processes of Christianisation and Islamisation are also an important force, although in a different manner in different parts of the continent. Economic and technological developments often also are driven by these religious developments, as well through migrant worker networks as in transformations on local levels. Further, the potential of different religious groups to impact society again depends on economic and political factualities that in themselves are not religious. The rise of societal influence of the Sufi-Islamic Mouride brotherhood in Senegal, e.g., could take place only in the political voids left by colonialism in their destruction of traditional kingdoms and nobility, as well as in the frame of economic opportunities created by the colonial trading system, and the new global trading networks that came after decolonisation (cf. Barry, 1988; Ross, 2006).

Taken from their own perspective, movements which focus on the purification of monotheistic religion, and/or the conversion of people, may coincide with the modern secular outlook which values technological and economic progress, overruling traditional practices based on the belief in the sacredness of nature. An example can be seen in how the Mouride brotherhood in Senegal, valuing prayer and work as central values, has played a pivotal role in cutting down several forests (which are traditionally seen as sacred places) in Senegal for purposes of peanut production.[3] All the same, whereas the Islamic theology in principle robs trees from the protection by guarding spirits (because the might of God/Allah will always be greater than that of any spiritual being), it still lends a spiritual meaning to nature in general, and to trees in particular, as the symbol of heavenly paradise that Allah holds for his believers[4] (Ross, 2006).

A framework that is also of importance for understanding the plural meanings of trees is the modern-secular framework that drives the work of many international environmentalist and conservationalist organisations working on the African continent. Such a framework will make us see religious movements, as well as traditional beliefs and practices, only as having a positive or negative effect on the thriving of the environment, without valuing their normative viewpoints in their own right. The environmentalist/conservationalist framework has its own normative views, of course, of which sustainability is a key one. Our ways to live in the world should be sustainable, that is, the human race should not use up resources, looking only for short-term profit, and creating long-term death and destruction. All the same, propagators of this framework never ask about the perspective expressed in the word 'environment.' Environment for whom? For the human race, in the end. Allthough striving for a sustainable environment for humans will probably have the effect that animals and trees and other living beings also have better chances than if we strive only for profit and domination of nature, the perspective of secularism can not ask about the meaning of nature, or of trees, for that matter, in a wider frame than that of humanity. Humanity cannot be understood, by secularists, from an angle in which non-human or supra-human values count.

Finally, there are also traditionalist movements, which, in an attempt to counter the negative influences of the colonial era, aim for a return to old religious beliefs and traditional social customs. An example is to be found in Benin, where Vodun religion has been reinstalled on a national level. Traditionalist movements have for their aim to preserve (or, if necessary, to reconstruct or reinvent) traditional knowledges and practices in which the relation of humans to their environment is considered from a perspective in which the spiritual aspect of all beings is taken into account. Humans in that perspective can live healthy and prosperous lives only if they know how to address the spiritual world that transcends their strictly human aims. In the present situation, where the forces of global economies and nation-based politics reign, traditionalism will have to negotiate its values with those of, most notably, international tourism and national culture (cf. Juhé-Beaulaton and Roussel, 2003) and, through these, also with conservationalism. We see this happening where the traditionalists' view of certain forests as sacred has now been taken up by the international conservation efforts of the UN by declaring those forests cultural heritage, naming them ICCAs: Indigenous Peoples' and Community Conserved Territories and Areas (www.cbd.int/pa/doc/ts64-case-studies/senegal-en.pdf). These negotiations, however, may have the effect that the return to traditional spirituality will transform into a matter of cultural folklore.

When we zoom in on the subject of trees as bearers of meaning in an African context, the different frameworks mentioned above overlap, interact and come into conflict. For my purposes, I want to question not only how these different frameworks prestructure any possible hermeneutics of trees, but also how the different meanings given to trees can enter into dialogue with each other. The meaning of trees should not be understood one-dimensionally, but we should carefully disentangle the different meanings bestowed by different groups in societies, and thus take their frameworks of reference into account.

Religion and the symbolism of trees

A first example of trees' contested meaning is provided in the case of the cutting of the Mbegué forest (Khelkom in Senegal), described by Schoonmaker Freudenberger (1991), where deforestation was carried out for economic and religious reasons (to provide opportunities for the agricultural labour through which salvation is supposed to be reached). In this case, we see how monotheistic religion can suspend possible cultural or spiritual inhibitions against felling old trees whilst at the same time stressing their symbolic religious meaning (i.e., the 'right' religious meaning). Schoonmaker Freudenberger has described in a passionate tone how in 1991 Mouride faithful, following a call of their religious leadership on the radio, cleared a whole forest in just a few weeks:

> From village and city, the faithful flocked to Mbegué, bringing axes and saws to clearcut 173 square miles of one of the last remaining forests in Senegal's degraded heartland. In three weekends, they felled more than five million Sahelian trees and shrubs. [. . .] For his part, the Khalifa-General (supreme head) of the powerful Mouride Islamic brotherhood was well on the way to meeting his goal: 45,000 hectares of newly cleared and fertile land would soon be put into peanut production.
>
> (Schoonmaker Freudenberger, 1991, p. 1)

In this example, we see a conflict of meaning given to trees by the secular environmentalist author (who wants to ward off desertification and uphold a livable human habitat) and the religious Mourides, who negotiated economic opportunities and a theology of the good religious life, resulting in the ending the life of the forest. Underneath these conflicting views there is also a conflict between two peoples/cultures: the Wolof Mouride faithful and the Fulbe herders who traditionally lived in the forest, and who were displaced by the (legal) cutting of the trees. The Fulbe possibly also hold more traditionalist views of the spiritual meaning of trees, although they just as well consider themselves Muslims.

A second example of the contested meaning of trees is from a story narrated to me by Michael Onyebuchi Eze on the tree called Uvuru, which was dominating the central square of his native village in the present day Enugu state of Nigeria. Ukwu Uvuru may have been more than 1,000 years old, according to reports, and had served from time immemorial as a place for public assemblies, as a source of fruits for children sitting in its shade, and it was ascribed a status of sacredness. It was cut down in 2002 by young Christians, who had been motivated to do so by their pastor. The motivation given was that the tree not just symbolised, but actually embodied, evil forces which would explain the social and economic problems of the locals. It supposedly held the people chained to the past and would hinder progress – so chopping it down was expected to solve that.

Although I don't know more details of the case, it seems a clear example of what happens when Christianity claims its ground by opposing traditional (folk) beliefs and practices. The idea is that ancient sacred places (or the opposite – places of evil), be they a watersource, a magnificent tree or a crossing of roads, should either be Christianised or be removed. In many places in Europe, you see images of the Christian cross, or road altars for certain saints, where there used to be places of ancient taboos, or of worship – like under an oak, at a well, or at a crossing.[5] Surveying the literature makes clear that traditional spirituality, which often survives in folk beliefs and practices, is seen as a problem in the eyes of many Christian leaders. In the example of Ukwu Uvuru, the loyalty of people towards anything traditional, even if there might not be an attitude involved that would contradict Christianity, is seen as a sign of lack of religious trust.

In religious studies, a sociological and anthropological point of view is predominant – leading to the explanation of the resurgence of traditional beliefs as a consequence of problems of modernisation (cf. Juhé-Beaulaton, 2008, p. 8, and Ter Haar, 1992, p. 111). Economic difficulties, or the alienation resulting from urbanisation, would be the main factor in seeking help in the spiritual realm of traditional deities. Christian theological discussions of the matter sometimes harmonise with these views in describing traditional spiritual practices as disturbing the order in society. We find an example of this approach in a recent thesis on the tensions between belief in deities and Christianity in contemporary Igbo culture. The author, Christopher Okwor, writes as a recommendation that

> The church should as a matter of necessity adopt a holistic approach to evangelisation. Christianity must be ready to feed the deep spiritual and material quests of the Igbo *in order to control them and be able to divert their attention from deities* (italics are mine). [. . .] Shrines are valuable heritage of our past. [. . .] Burning of shrines should be treated as a very serious offence. [. . .] *Shrines should be developed into tourist centres. They are homes for*

> *endangered species of plants and animals and some have beautiful caves and springs.*
>
> (Okwor, 2012, pp. 134–135)

The central problem with traditional practices of worship of deities seems to be that they escape the dominant morality and legality and are therefore 'beyond control.' Here we see the secularist framework, with its two main interests, tourism and sustainability, join hands with a Christianising interest: the striving for a moral life according to biblical prescriptions for the good life. We see the same effect in the work of Juhé-Beaulaton, a historian, who directs attention to issues concerning law and order surrounding sacred forests and shrines. Like Okwor, she sees a harmonisation between conservationalist, traditionalist and the legal and economic interests of the state as a solution to preserve the 'beautiful' places of worship whilst pacifying the spiritual needs of the people (Juhé-Beaulaton, 2008).

In these approaches, however, the inherent value of the spiritual meaning of these shrines, often part of small forests, and therefore of the trees that mark and localise them, is lost, thus excluding the other mentioned framework which claims certain meanings for the environment in general, and trees in particular, namely the framework of a more spiritual outlook on things, in monotheist as well as in traditional religions. More apprehension for this aspect is found in the work of geographer and Islamologist Eric Ross, who has studied the symbolic meanings and usage of trees in West Africa, especially in Senegal, in their own right. While in non-African contexts the central tree in an African community has become known as "palaver tree" (from the Portuguese word for speech or discussion), Ross indicates that this name is too narrow:

> [it] designates what is in reality a number of different phenomena which make political, social or religious use of individualized trees. [. . .] Rather than a single 'palaver tree' serving as locus of public debate, polities were marked by a number of different trees, of various species, which served a variety of public and collective functions, only one of which was the 'palaver' process.
>
> (Ross, 2008, p. 136)

Through field research, Ross has located and described many ancient trees in Senegal, which often have survived the royal palaces or the villages of which they at one time were the centre. He has opened up a new branch of research, combining the locating on google maps of the individual trees, photographing them, and describing the oral histories that are told by the people living there. Thus he has made a beginning to make local knowledge accessible to the wider world: the knowledge that distinguishes between the public, social, political, and religious functions of trees. The ancient, hollow, baobabs are known to have been used as shrines, altars, and tombs for the griots (court singers, like the medieval European 'jesters' or 'troubadours'). Trees are also mentioned to localise where a battle took place, or a boundary existed (Ross, 2008, pp. 136–137). In their function as the embodiment of political legitimacy, they can be compared with the obelisks, triumphal arches, and other symbolic structures in Europe. In addition, trees could be the places where justice was rendered. In these public functions, they were and are seen as 'places of power' and places of memory (Ross, 2008, p. 139 and p. 144). Trees thus function as

> [. . .] markers, as memorials and as monuments. These functions are spatial in that they contribute meaning to the landscape, but they are also social and political, in that they 'fix' identities while also articulating a spiritual worldview.
>
> (Ross, 2008, p. 146)

Trees as archetype

In his work on the sacred city of Touba (an elaborated and reworked version of his dissertation), the urban place which is the centre of worship in Senegalese mouride Islam, Ross explains the mystical and symbolic understanding of trees. Mouride mysticism is a modern branch (defying the conflict between modern and traditional) of neoplatonic Sufi understanding of the world in relation to God as creator and provider of grace for human beings. The city of Touba, in the inner land of Senegal, was founded in 1887 by Shaykh Ahmandou Bamba Mbacké (1853–1927), on the authority of a revelation from God about the locality of this sacred place. According to a local legend:

> After a patient search his (of the Shaykh, AR) attention was drawn, one day, to a tree which stood out clearly from the others by its size, its importance and its peculiar location. The tree [. . .] stood on a plateau on the spot where the dome of Touba Mosque now stands. Ahmadou Bamba prayed in the shade of the tree, and that is where he had his long awaited revelation.
>
> (cited from Ross, 2006, p. 28)

The sacredness of Touba is experienced and reinforced in the yearly pilgrimage believers make to the city, where they offer prayers in its grandiose mosque, longing to be buried within the city when they die, which for them symbolises the closest connection to heavenly paradise. Touba, the name of the city, originally comes from the word tûbâ, which means the Tree of Paradise in Islamic tradition. It stands for the bliss of the heavenly state of the righteous after death.

As trees often play a critical role in foundation legends of places of human habitation, as well as of religious movements (think of Buddha, who also received a revelation under a tree, and the symbolic 'tree' of the Christian cross), Ross explores the specifics of this role in the case of Touba within the context of Sufi neoplatonic understanding. Here Touba is seen as "a qutb, a 'pole' or axis mundi. [. . .] The concept of qutb is primarily an astronomical one. It designates the 'celestial pole,' a hypothetical spot in the sky around which the heavens revolve" (Ross, 2006, p. 18). In neoplatonism, the cosmos is understood as an unfolding of reality in layers of being, emanating from God, which are closer or farther removed from his, so to speak, ontological and spiritual gravity centre. According to Ross,

> Sufism has invested this astronomical term with several related spiritual meanings. The term qutb is used to describe *transcendence*. It can be applied to any being, moment, event, or place which connects various layers of reality to each other.
>
> (Ross, 2006, p. 18)

It can be a moment, which thus should be remembered with reverence – like the moment of revelation of an individual. It can also be a special, rare and beautiful place, such as a river, a grove, a high mountain, or a city (cf. Jerusalem, which is understood as a place where the divine can be almost 'touched' by Jews, Christians, and Muslims). It can also be, according to Sufism, a person – which then leads to taking, for instance, the grave of this person as a place for pilgrimage. Touching the grave or some other act of dedication can then help the believer on his/her spiritual path. Ross further explains this in more philosophical language by citing Mircea Eliade:

> A sacred place constitutes a break in the homogeneity of space; this break is symbolised by an opening by which passage from one cosmic region to another is made possible (from heaven to earth and vice versa; from earth to the underworld); communication with heaven is expressed by one or another of certain images, all of which refer to the axis mundi, pillar [. . .], ladder (cf. Jacob's ladder), mountain, tree, vine, etc; around this cosmic axis lies the world (= our world), hence the axis is located 'in the middle', at the 'navel of the earth'; it is the Center of the World.
>
> (cited from Ross, 2006, p. 20)

The qutb, the axis, can therefore also be a tree, and it often is. To make things more clear for modern readers, Ross introduces the term 'archetype,' coined by Carl-Gustav Jung in his psychology of the collective subconscious. He makes clear that one does not have to accept Jungian drive-psychology to use this concept, which just indicates what Sufi theosophism means by fixed essences, the forms which 'hold,' or refer to, the divine presence of God. Although, as is clear from the above, designating trees as archetype of the relation of humanity to the spiritual realm is in no way restricted to an African context, trees play an important role in African spiritual symbolism (next to all kinds of other signs, of course, such as water, certain animals, fire, etc.).

As Ross's research into the 'palaver' trees has made clear, the central tree of a village or royal courtyard often survives the mud-structures of the human habitats themselves. The trees thus serve as memorials of human deeds and experiences in the past. In the specific instance of Touba, the belief in the spiritual tree, symbolising the entrance of paradise, is considered as the "celestial register upon which the names and deeds of individuals are recorded" (Ross, 2006, p. 31). The symbolism of the tree is even more elaborate, as the words of Cheickh Tidiane Sy make clear:

> In Islamic tradition, Touba also designates a tree of Paradise on whose leaves are inscribed each human's good and evil acts. Each leaf, as it falls, inexorably provokes the death of the individual whose acts have been recorded. The leaf is then preserved for the Day of Judgment.
>
> (cited in Ross, 2006, p. 32)

To inscribe themselves on this divine spiritual record, Mouride believers scratch their names and those of their loved ones on an actual Baobab tree in the city of Touba. In this religious practice, there is thus no sharp distinction between a material act (taking out a knife and carving letters into the bark of a tree, like lovers do in modern Europe, and perhaps all over the world, to pledge the eternity of their bond) and a spiritual act of faith (believing in or hoping for the saving grace of paradise, which is just another metaphor for the nearness of God).

This 'mixing' of matter and mind, of belief and practice, shows how 'shamanistic'[6] spiritual ontologies are present in monotheistic religion, even if full shamanistic ontologies do not focus so much on unity, as in the unity of God, but see all phenomena as potential manifestations of spirit, or as vehicles to reach the spiritual realm. Moving in the spaces of such 'hybrid' (to borrow a concept of Bruno Latour) realities is contrary to the protestant (modern Christian) idea that only belief can save souls, for which reason much of traditional ritual behaviour then should be considered irrational, and even banned. In this respect radical Protestantism reflects the same spirit as secular modernism (although I would not want to make a historical claim as to one inspiring the other). And since this secularism forms the actual (although not the necessary) framework of most modern philosophy, it has become hard in philosophy to understand archetypal relations of humans to trees as something more than irrationality.

The relation to ancient mysticism, which is so central in Mouride Sufism, can be put in a wider framework, as Ross has done in an article from 1994. He there tries to understand present day Islam from an Afrocentric perspective. In Afrocentric research into the origins of African civilisations (originating in the work of Cheikh Anta Diop, and leading up to present day writers such as Sarwat Anis Al-Assiouty), the link with ancient Egyptian cosmology is more and more researched, as well as its influencing role in Greek, Semitic, and, in the end, European and modern African science and religion. From a more philosophical perspective, this relation resonates in the seminal work by V. Y. Mudimbe, *The Invention of Africa*, which, in a productive dialogue with the philosophy of the French philosopher Foucault, develops the understanding of African philosophy as *gnosis*, the neoplatonic word for higher wisdom, which can be sought by means of the secret training of the initiated. This concept of gnosis, then, according to Mudimbe, is also a vehicle to transcend the cultural-anthropological concept of 'Africa' as a special place, or to transcend the localising concept without leaving the place. Even to transcend the discipline and overcoming its colonialist preconditions by making it into "a more credible anthropou-logos, that is a discourse on a human being" (Mudimbe, 1988, p. 186).

Trees – contested meanings

After outlining the attitude towards the meanings of trees in the crossroads of conversion movements, traditionalism and modernisation, we still have to take a closer look at the secularist point of view before we can try to show how the different frameworks might be brought into dialogue with each other.

The secular perspective, which is central to most environmentalism as well as to the investigation undergirding it, shows very clearly in the reports of Schoonmaker Freudenberger cited above, in the second section. In her account of the cutting of the Mbegué forest we see a shrill contrast with the tree-friendly mysticism of the Mouride brotherhood described by Ross. Schoonmaker Freudenberger shows that she has a more positive idea about the Fulbe pastoralists, who were displaced by the cutting of the forest, and who have been in conflict with agriculturalists in West-Africa several times. This choice is based, not on a more favourable view on one group's spiritual outlook or way of life, but on insights from environmental science and prospective expectations of which lifestyle in the Sahelian landscape might be more sustainable.

It is of importance here to underline the positive aim of secularist environmentalism – to preserve the earth for future generations of human beings, as a beautiful and healthy place to live, among a wealth of other species – plants and animals. From this aim for the future, environmentalism lends its legitimacy to critically question the behaviour of human groups, peoples, but also governments and the corporate world. Its voice is an important one in the shared human efforts to live a good life, and its adherence to planning on the basis of scientific research (objective data) provides an indispensable complementary discourse to the religious conversion movements described above. All the same, one has to be critical towards its accounts too – firstly, because of the provisionality of all scientific results, which often rest on researching a limited amount of factors, leaving others in the dark, and, secondly, because it fails to grasp the role symbolism and spiritual meanings may play in motivating human behaviour. If we put this second point to work with respect to the double-edged sword of Mouride theology – respecting individual trees as representative of the spiritual tree of paradise, whilst also stimulating agriculture as a good way of life for believers, even though it kills actual trees – we might be able to not just criticise the role of the Mouride leadership, as Schoonmaker Freudenberger does, for having too much power in a relatively weak state, but we might criticise it on a theological level itself.

What would such a theological criticism look like? It should direct itself at the level of understanding the relationship between the visible/material world and the religiously understood spiritual world (the paradise of God, the afterworld). In Senegalese Mouridism, there might be a conflict at work between a 'shamanistic' ontology, which considers matter and spirit as potentially interacting, and a dualist ontology of heaven and earth being separate places. This kind of conflict is not specific to this branch of Islam, but can be seen in monotheistic traditions in general, and even in modern secularist worldviews where they deal with the relationships of human beings to things. Whereas shamanistic ontologies recognise each (living) being as an expression of (divine) spirit, monotheist ontologies attribute a unique meaning to the human being as the servant of God, playing a role in his plans with the cosmos. The role the human being plays in monotheism is understood to take place in two realms simultaneously: s/he can live a moral life in the visible/material world, actually helping others or making the world a better place to live in, but the real meaning of these actions lies in the 'other' world, the world of eschatological issues concerning the spiritual salvation of the human and/or of everything created.

We find an example of this monotheistic approach in the Mouride felling of the Mbegué forest, as it lets the aims of spreading the faith and the striving of humans to deserve heavenly bliss in the afterlife prevail over the life of the trees concerned. All the same we saw the meaning of trees not to be absent in Mouride theology – as it is preserved symbolically in the reverence for certain exemplary individuals in specific holy places.[7] Here we see a reminder of how shamanistic ontology can live in monotheistic religions, not as a strange element to be rooted out (like reformation movements such as radical protestantism or salafism aim to do), but providing ways for religiosity to be experienced and acted out in human lives. Would this shamanistic element be more explicitly acknowledged, it might be used to criticise and correct violent acts of religions against our natural surroundings, harmonising traditionalism and modern religiosity. In such an acknowledgement, we can see an alternative to the strictly secularist discourse of environmentalism, while creating the terms for a dialogue on the religious interpretation of our environment.

Conclusion

To conclude I will discuss a philosophical attempt to take the spiritual and the material wellbeing of human beings in relation to trees simultaneously into account, as presented in an article by Michael Onyebuchi Eze,[8] which aims to provide an Africanist theory of environmental ethics. His holistic viewpoint aligns with descriptions of 'African' ontologies as given by the Belgian Franciscan Tempels and Kenyan theologian/philosopher John Mbiti. Both authors worked to present a view of African religion, ontology, and ethics in general, in reaction to ethnographic work from the colonial age that provided descriptions of so many different African cultures as local and exotic – over and against (Western) universal religion and ontology. (cf. Ellis and Ter Haar, 2009, p. 404) In Tempels' work, the concept of life force is especially important. Tempels summarises the 'Bantu' metaphysics as an ontology that encompasses all being and understands it as interdependent. Whereas *Bantu Philosophy* claims to reconstruct reality as it is, it admits this reality to be accessible through different cultural frames in different ways. Following Tempels, Eze coins an understanding of ontology as dependent on cultural frames on the one hand, but as simultaneously providing a *true* description of things on the other, even if it may be mythological.

> It does not matter whether the exploits of Api or Uhere (river deities who were considered to protect the people, AR) during the Nigerian civil war is true or false. As a myth or legend however, what it does convey are certain philosophical truths [. . .] of the people's cosmology and understanding of their environment. Their understanding and relationship to it is adapted through this socio-cultural mindset.
>
> (2017, in print)

Eze also quotes Ramose in this context:

> According to Ramose, this idea of community includes 'the greater environing wholeness in the sense of both the encompassing physical and metaphysical universe, together with the human universe in the sense of community.' This community is not just a collective of humans, it is a fluid habitation of interactive forces, beings, elements, animate and inanimate matters of the environment.
>
> (2017, in print)

For our purpose, to understand the meaning of trees, the most important conclusion of Eze's article is that 'environment' cannot be understood as the counterpart of an independent humanity. Nor can the aim of religion be taken to be opposed to a nature devoid of spirit. "The sacredness of nature is because all elements have vital force." This leads then to the insight that "The environment is not just inconsequential, it is part of life and constitutive of humanity." In this view, then, our relationship to trees should be caring and nurturing rather than economic, as the trees, together with all of nature, deserve of a respect as bearing the possibility of human life. Here we have a view that reconciles philosophically the aims of environmentalism and a certain spirituality which encompasses respect and prudence in human-nature relations.

Although I agree with the ideals as worded by Eze, his philosophical underpinning of them fails to disentangle the practical difficulties which arise out of the misunderstandings between the factual and different frameworks indicated in this work. It is my conviction that only by articulating the frameworks, highlighting their loci of conflict and internal paradoxicality, can we work towards a richer hermeneutics of trees (and the environment in general), that can form the basis for the dialogue humanity has to carry out – a dialogue that will try to reconcile the insights from science, from religious theology, and from spiritual practice, and bring them to fruition for a better cohabitation of different species on this earth. In my view, we need a multi-dimensional understanding of the meaning of trees, which alone will be able to provide a sustainable basis for conservationist policies – an understanding that will take a) scientific research, b) religious movements and theologies, and c) traditional spirituality and healing rituals serious. African philosophy as well as historical, geographical, and religious knowledge developed on the African continent provide, as I have shown, important sources for such an understanding, as well as for the envisioned dialogue from which, first, African regions and cultures, but, in the second instance, also those beyond Africa, might benefit.

Notes

1 I want to express my gratitude to Michael Onyebuchi Eze for letting me read and make reference to an article which is still in press, as well as for commenting to the first version of this paper.
2 In Roothaan, 2015 I have argued against restricting the concepts 'shamanism' and 'shamanistic' to the (practices of) Siberian spiritual mediators from whom it originally stems, and for following I. M. Lewis where he states: "I prefer to see shamanism as a general, cross-cultural phenomenon based on the shaman's mastery of spirits and the practice of this art with the aid of spirits." When we broaden his wording by substituting 'the spiritual' for 'spirits,' it is clear that the concept also covers certain monotheistic, buddhist, and agnostic practices.
3 A branch of agriculture originally stimulated by the colonial trading possibilities, although later colonial authorities worried about the desertification that might be the result – cf. Schoonmaker Freudenberger, 1991, pp. 7–8 and Schoonmaker Freudenberger, 1993, p. 15.
4 As several authors have shown, however, monotheistic conversion/purification movements also often vilify traditional spiritualism, by calling its practices immoral or amoral (cf. Ellis and Ter Haar, 2009 and Ter Haar, 1992).
5 The oak is almost the only tree that is still known to the public in Europe as symbolising a sacred place in pre-Christian times. But of course, the pre-Christian peoples knew all the trees for their specific natural and spiritual goods and evils. Similarly to the pre-Christian traditions of Europe (of which some living practices can still be found in remote rural areas), traditional African societies know/knew the different species of trees for their medicinal usage, their food, and attribute(d) them also religious or public functions.
6 For a discussion of the word 'shamanistic' and why I choose to use it, see Roothaan, 2015, and footnote 2.
7 In Roothaan, 2015 I have shown how monotheist founding narratives are rooted in shamanistic imagination – which might put in question whether the new religions in fact replaced the old ones as is commonly thought, or rather adapted them to new lifestyles.
8 Since this article is still in production, no page references can be given.

References

Barry, B. (1998 [1988]). *Senegambia and the Atlantic Slave Trade*. Cambridge, New York and Melbourne: Cambridge University Press.

Cabral, A. (2007 [1979]). *Unity & Struggle. Speeches and Writings* (2nd ed.). Unisa and Hollywood: Unisa Press and Tsehai Publishers.

Ellis, S., and Ter Haar, G. (2009). The Occult Does Not Exist: A Response to Terence Ranger. *Africa*, 79(3), 399–411.

Eze, M. O. (2018). Humanitatis-Eco (Eco-Humanism): An African Environmental Theory. In *Handbook on African Philosophy*. New York: Palgrave Macmillan.

Juhé-Beaulaton, D. (2008). Sacred Forests and the Global Challenge of Biodiversity Conservation: The Case of Benin and Togo. *Journal for the Study of Religion, Nature and Culture*, 2(3), 351–372.

Juhé-Beaulaton, D., and Roussel, B. (2003). May Vodun Sacred Spaces Be Considered as a Natural Patrimony? In T. Gardner and D. Moritz (Eds.), *Creating and Representing Sacred Spaces* (pp. 33–56). Göttingen: Peust & Gutschmidt Verlag.

Mudimbe, V. Y. (1988). *The Invention of Africa: Gnosis, Philosophy and the Order of Knowledge*. Bloomington, IN: Indiana University Press.

Okwor, C. O. (rev.) (2012). *The Significance of Deities in the Contemporary Nsukka Northern Igboland.* Master's thesis University of Nigeria, Nsukka. www.google.nl/url?sa=t&rct=j&q=&esrc=s&source=web&cd=4&cad=rja&uact=8&ved=0ahUKEwjur72KxNfOAhUJAxoKHQq9AN4QFggyMAM&url=http%3A%2F%2Frepository.unn.edu.ng%3A8080%2Fjspui%2Fbitstream%2F123456789%2F771%2F1%2FOKWOR%2C%2520CHRISTOPHER%2520OKONKWO.%2520_REV.pdf&usg=AFQjCNELn7jDfDuLzpRUBtb1LJ5YNX481A&sig2=R9t_yyUJQ9ldl3XOJdfOEQ (accessed August 22, 2016).

Roothaan, A. (2015). The 'Shamanic' Travels of Jesus and Muhammad: Cross-Cultural and Transcultural Understandings of Religious Experience. *American Journal of Theology and Philosophy*, 36(2), 140–153.

Ross, E. (2006). *Sufi City: Urban Design and Archetypes in Touba*. Rochester: University of Rochester Press.

Ross, E. S. (2008). Palaver Trees Reconsidered in the Senegalese Landscape. In M. J. Sheridan and C. Nyamweru (Eds.), *African Sacred Groves: Ecological Dynamics & Social Change*. Athens, OH: Ohio University Press.

Schoonmaker Freudenberger, K. (1991). Mbegué: The Disingenuous Destruction of a Sahelian Forest. *Paper No. 29*, International Institution for Environment and Development.

Schoonmaker Freudenberger, K., and Schoonmaker Freudenberger, M. (1993). Pastoralism in Peril: Pressures on Grazing Land in Senegal. *Drylands Program: Pastoral Land Tenure Series No. 4*, International Institution for Environment and Development.

11 Global warming as an ontological boomerang effect

Towards a philosophical rescue from the African place

Victor C. A. Nweke

Introduction

This chapter is specifically tailored to address the challenge of global warming from the perspective of some theories in African philosophy as an academic discipline. The practice of philosophy in any place involves critical reflection on human challenges. When we study the history of philosophy, we seek to learn from the past philosophers in order to confront our own challenges. For centuries, *Western philosophy* was globally projected and taught in institutions of learning as *philosophy*. One of the effects of this is that it inflicted most non-westerners, the recipients of Western education, with what Catherine Acholonu (1995) calls "intellectual blindness," which makes it very difficult for them to realise, explore and adapt the ideas of philosophers in non-Western places to confront the challenges of their place. It took years of polemical debate for African philosophy to be recognised as a legitimate philosophical enterprise. A basic heritage of the prolonged exclusive preoccupation with Western philosophy across the globe and its subtle contemporary dominance today is the underexploration of the intellectual heritage of Africa (Metz, 2016a, p. 171).

Consequently, the challenge to experts in African philosophy today is to explore, mine, articulate, inaugurate and project novel thoughts from the African philosophical place geared towards the Global Expansion of Thought[1], i.e., for addressing human challenges in and beyond Africa. In this connection, the major objective of this chapter is to address the challenge of global warming from the African place. In line with the aim of the entire book, this is done with the hope that it will help to reduce the rate of the "intellectual blindness" that makes it very easy for intellectuals to readily accept Western paradigms but to hesitate to seek, explore and accept viable non-Western paradigms as legitimate (Acholonu, 1995, p. 1). In pursuit of this objective, I shall undertake a critical and creative conversation with the works of scholars in African philosophy as well as the ideas of anonymous African philosophical sages encapsulated in African oratures.

Generally, the chapter is divided into three major sub-headings. In the first, I explicate the general perception of global warming (GW) and its challenge in the contemporary world. The second takes a look at the challenge of GW from the standpoint of certain theories in African philosophy. I use relevant principles and theories articulated by Innocent Asouzu, Jonathan Chimakonam and Thaddeus Metz to explain the basic cause(s) and possible ways to tame the tide of GW. My approach in this regard is conversational rather expository in the sense that I interrogate and unveil what I take to be the prospects and limitations of each of the theories. In the last sub-heading, I indicate the inherent relationship between the theories examined and their prospects in relation to understanding and addressing the challenge of GW. I specifically submit that each of the theories examined contains plausible ideas that can be creatively harnessed to provide a philosophical framework for understanding and reversing the tide of global warming today. I then conclude with a call for a creative synthesis of the ideas into a coherent Africa-inspired philosophical framework for sustainable development.

The reality and challenge of global warming (GW)

Global warming has to do with measurable increases in the average temperature of Earth's atmosphere, oceans and landmasses. Scientific research shows that the earth's temperature has never been stable but that the current rapid warming of the earth is unprecedented in history. At first, the scientific community was sceptical whether the increased rate of GW in contemporary times is necessarily dangerous. Some policy makers were also indifferent to the question of GW. This is no longer the case today. GW is indeed one of the most debated science issues of the 21st century that challenges the very structure of our intricately interconnected world. It is "not just a scientific concern, but encompasses economics, sociology, geopolitics, local politics, and individuals' choice of lifestyle" (Maslin, 2004, p. 1). The reason is simple; the current tide of GW is not just uncomfortable but life threatening. It is believed to be causing severe damage to the environment, humans and all other creatures in it (Reay, 2006).

The threat of GW to the environment and biodiversity, including humans, may now be regarded as a reality rather than just a prediction. Its effects are very evident and visible, including the increase in extreme weather events, rising sea levels, disappearing glaciers and polar ice, damaged coral reefs, changes in wildlife distributions and health, and increased human activity and abundance of disease vectors (Wang and Chameides, 2005). The scientific community also predicts that the challenge of global warming will "increase over the next several decades in the absence of radical measures" (Metz, 2016b, p. 116). More specifically, Maslin admonishes: "Be under no illusion: if global warming is not taken seriously, it will be the poorest people in our global community, as usual, that suffer most" (2004, p. 3). This point has also been corroborated by scholars such as Metz (2016b). As the dangers of GW have become more evident, the scientific community has intensified their research to ascertain the basic factor that is responsible for the present tide of GW. The wide consensus among existing research findings is that it is not a natural phenomenon. The rate of global warming we experience now is said to be a result of human activity.[2] There is also a widely shared opinion "that our modern civilization, science and technology have not the right *disposition*, or even the weapon to deal with the problem, we [therefore] have no choice but to seek alternatives" (Acholonu, 1995, p. 5). As an indigenous African, and a propagator of African intellectual heritage, I have chosen to seek an alternative philosophical approach to GW from the African place.

Addressing the challenge of global warming (GW) using theories in African philosophy

There are many existing and emergent principles and theories in contemporary African philosophy. I will here employ Asouzu's principle of integration, Chimakonam's theories of African science and conversational thinking, and Metz's theory of relational ethics. My choice of theories should not be construed to mean that these are the only theories in African philosophy for addressing GW. Nonetheless, they are, to my knowledge, the most attractive metaphysical, epistemological, methodological, scientific, and ethical theories in African thought that can serve as the building blocks for the type of Africa-inspired philosophical framework that I propose. I, therefore, proceed to address the challenge of GW from the perspective of these theories.

Asouzu's principle of integration and the challenge of GW

It has been argued that African philosophy as a systematic discipline has one of its most articulate expressions in Asouzu's Complementary Reflection (ACR).[3] The basic ontological principle that undergirds ACR is the principle of integration. This principle simply states that "anything that exists serves a missing-link of reality." By missing-links, Asouzu refers to everything that exists, animate and inanimate, material and non-material (2004, pp. 277–278). What this means is that to be is to be a missing-link of reality. Being is therefore a conglomeration of all existents (Edet, 2011, pp. 28–32). The principle of integration simply affirms that all existents are necessary entities that play an essential role for the joy of every being in the universe. The wellbeing of every reality in the universe is so intricately interdependent that the constant careless abuse of the wellbeing of a particular entity in order to promote the wellbeing of another entity often leads to negative results. The consequences that arise from any fragrant abuse of the principle of integration are what Asouzu describes as an *ontological boomerang effect*. A careful study of the present tide of GW suggests that it is an ontological boomerang effect of the careless disruption of the principle of integration by humans as they seek to promote their wellbeing. An exposition of the prevalent way humans see, approach, and exploit nature to meet their needs will explicate this point.

The most utilised tools for the promotion of human wellbeing in human history are products of modern science and technology (MST). R. T. De George notes that: "In its common Western sense, modern science is the study of nature and its laws; technology is the application of science and the laws of nature to specific human projects" (1994, p. 15). This common Western sense of MST, is perhaps, the universal commonsense conception of MST. Nonetheless, the theoretical foundation and methodological dispositions of MST are founded on, and are driven by, a given conception of nature that is peculiar to Western philosophy. De George identifies three basic features of the peculiar Western philosophical conception of nature that drive MST. The first is that "nature is a material reality opposed to and different from mind and spirit" (DeGeorge, 1994, p. 15). This view of nature favours mechanistic rather than teleological explanations of natural phenomena. Second, nature is seen as "an *object* for thinking subjects. . . . For it [nature] is there for humans to interrogate, subdue, dominate and manipulate" (De George pp. 15–16). The last view is that "nature is value free." In other words, there is no room for value judgments (De George p. 16). Sequel to these three interrelated views, De George submits that:

> Western philosophy has tended to give up natural law and to treat human beings as the measure of all things. Ethics, or the study of how human beings should act, comes not from the scientific study of nature but from human beings.
>
> (p. 17)

Consequently, one will understand that MST is ultimately based on certain fundamental presuppositions of Western philosophy (little wonder, modern science is often described as Western science). Nonetheless, it is pertinent to note that the prevailing disjunctive mode of Western philosophy is actually articulated, consolidated, canonised and projected as the sole legitimate paradigm by androcentric and ethnocentric Western male philosophers for many centuries. Technically, Asouzu (2011) calls this disjunctive logical mode of Western philosophy "the philosophy of essence." He also insists that it all started with Aristotle's first philosophy. It is specifically a philosophy that bifurcates realities into the essential and the accidental; the indispensable and the dispensable. Whereas the former is indispensable and necessary, the latter is dispensable and negligible. Riding on the crest of this view, MST sees humans as the essential and indispensable entity in the universe. Every other entity is an accidental and dispensable entity that can be conquered, subdued, dominated, manipulated, or exploited for human wellbeing. Fidelity to this proposition necessarily disrupts the principle of integration. Hence, the challenge of GW is essentially an ontological boomerang effect of the incessant abuse of the principle of integration by humans for centuries. The question here is this: Given that the quest for human wellbeing in the contemporary world is driven by MST, and MST, as some have argued, is somewhat responsible for GW, how possible is it that humans can overcome the challenge of GW without denying themselves the benefits of MST? Is it possible to redefine the theoretical presuppositions and methodological dispositions of MST in a manner that MST will still be able to promote human wellbeing without the negative effect of GW? Chimakonam's theories seem to give a positive answer to these questions.

Chimakonam's theories (African science and conversational thinking) and the challenge of GW

Subscribing to the thesis that strict fidelity to a disjunctive logical/methodological disposition disrupts the principle of integration and invites counterproductive results, Chimakonam (2012) calls for the articulation and adoption of an appropriate alternative scientific disposition. In line with the title of his book, *Introducing African Science: Systematic and Philosophical Approach*, Chimakonam (2012) defines, argues, and introduces the basic principles of African science as an alternative to the protocols of Western science.[4] According to him, Western science can be defined as "a body of knowledge whose pursuit is tied to the principle of empirical, testable and demonstrable protocol," while African science can be defined as "a body of organized knowledge concerned with enquiries into all shades of reality supported with rational explanations" (p. 5). He also explains that the basic difference between Western science and African science is mainly the scope and methodological disposition. Western science deals only with a specific type of reality, the empirical. It is strictly concerned with that which is empirically testable and mathematically demonstrable. It is therefore restrictive. In contrast to this, African science is concerned with the whole of reality: "All sides of reality are covered provided the methods of enquiry offer explanations and results that are rationally tenable" (Chimakonam p. 5).

Against this backdrop, the elucidations of Chimakonam (2012) on the nature of Western science are fundamentally in tandem with Asouzu's position that MST is a product of a bifurcated ontology and exclusivist epistemology which operates with a disjunctive logical/methodological disposition that is insensitive to the necessary complementary interdependence of all existents in the universe. He also accepts that the predominant perception in African philosophy about the basic nature of metaphysics and epistemology is comprehensive, complementary, critical, and non-absolute. This forms the basic philosophical presuppositions from where Chimakonam articulated the basic laws, principles, and methodological disposition of African science. He specifically posits that African science sees the universe as "a system" comprising of everything, including humans and all non-human animals and inanimate entities existing in all possible worlds.[5] And as a system, the universe strives on the *law of uniformity* (p. 48). One can explore and exploit any existent in the universe to advantage without incurring any negative effects if and only if he/she does so without violating the *law of uniformity*. Explaining this law, Chimakonam writes:

> The law of uniformity stipulates that nature is a system with different aspects all of which share uniformity and non-uniformity among themselves. Every aspect is to be combined with the ones uniform to it and non-uniform aspects are not to be combined. To do the latter would be to set nature at war with itself. This is what the so-called sophisticated western technologies do.
>
> (p. 49)

The point on prominent relief here is that there is a necessary link that interconnects all existents and any serious disruption of the necessary link between any of the existents in the universe often affects the wellbeing as well as the sustainable development of all existents in the universe. What this comes to is that an appropriate disposition towards ensuring the wellbeing of humans in the world must respect the law of uniformity. Consequently, the quest of MST to "conquer nature," instead of preserving it, is inconsistent with the law of uniformity. This quest makes MST to endorse the careless abuse of the law of uniformity in search of great technological innovations that often come with "dangers two times greater" (Chimakonam, 2012, p. 48). Buttressing this point, Chimakonam avers:

> Francis Bacon the father of modern western science inaugurated this rough and shove attitude of western science. He had set the creed: "let us put nature in a rack and compel her bear witness against herself". All the attitudes of a western scientist is summarized in the above. These attitudes can be grouped into two: those aimed at making war with (conquering) nature and those aimed at forcing the scientific truth out of the conquered nature.
>
> (p. 48)

Unlike the hegemonic approach endorsed by MST, African science endorses a complementary approach that seeks to study, understand, and exploit nature to advantage without any serious disruption of the law of uniformity. It is a theory of science that rides on the assumptions that no position is absolute, no truth is eternal and there are many ways for solving a given problem. The only supreme law that should never be flagrantly disobeyed is the law of uniformity (Chimakonam, 2012, p. 49).

On the other hand, Chimakonam's theory of African science has a philosophical base in his theory of conversational thinking. In it, every standpoint is questionable and the incessant critical and creative interrogation of any existing standpoint often leads to alternative discoveries for meeting human needs. This view is what constitutes Chimakonam's theory known as *conversational thinking* (Chimakonam, 2015a, 2015b, 2015c).

Conversational thinking refers to the methodological disposition of conversational philosophy articulated by Chimakonam and endorsed by the Conversational School of Philosophy (CSP).[6] Conversational philosophy is the major orientation in the New Era or Contemporary Period of African philosophy (Chimakonam, 2015c, pp. 11–12). It is characterised by the creative adaptation of the relevant postulations of both the traditionalist/particularist and modernist/universalist schools of African philosophy towards the systematic construction of novel ideas, principles, and concepts in a manner that "clearly defines African philosophy as a 'platial' enterprise commanding phenomenological preoccupations that do not lead to ethnophilosophy or strict metaphilosophy" (Chimakonam, 2015a, p. 464). My submission here is that Chimakonam's theory of African science rides on his idea of conversational thinking and the methodological disposition of his theory of conversational thinking is comprehensive, complementary, and context-sensitive.

Thus, it is a promising alternative methodological disposition for an environment friendly science and technology (EFT).

So far, Chimakonam's theories of African science and conversational thinking repudiate the hegemonic positions and methodological disposition of MST that project man as the measure of all things. However, they fail to provide an articulate ethical principle that should guide human relationship with other existents in the universe. This is a limitation because the challenge of global warming is also an ethical challenge.

Metzian relational ethics and the challenge of GW

By Metzian relational ethics, I specifically refer to the Afro-communitarian moral theory formulated by Thaddeus Metz (2007) from the moral values salient in sub-Saharan African societies. Notably, Metz's theory has attracted some commendations and very serious criticisms from revered African philosophers such as Ramose (2007). Our interest here is in the basic principle of Metzian relational ethics and its prospects in respect to the challenge of global warming.

According to Metz, a rough summary of the cardinal principle of his version of Africa-inspired relational moral theory is that "an act is right if it respects others in virtue of their natural capacity to commune or harmonize, i.e. to be party to relationships of identity and solidarity with others; otherwise, an act is wrong, and especially insofar as it prizes discordance" (Jones and Metz, 2015, p. 539). The emphasis here is on harmonious relationships. What this comes to is that an act is right if it promotes harmony and wrong if it promotes disharmony. Also, identifying with and/or assisting others to meet their basic needs in a manner that does not deprive them of their inherent dignity is a moral obligation (Metz, 2007, 2016a). Therefore, promoting a harmonious relationship and benevolence among the entities in nature is a moral obligation.

Invariably, by prizing harmonious relationships as a moral obligation, Metz's theory encapsulates the view that fidelity to the principle of integration is right. It will also condemn any human activity, including scientific research and innovations that "prizes discordance" over harmony. More so, Metz's theory will readily endorse any project that seeks to promote harmonious relationships as right. Mark Tschaepe notes that the cardinal principle of Metz's theory basically "supplies the humanistic moral rule that underlies the obligation the person has to the community and provides an understanding that this moral rule works to sustain the harmony of the community" (2013, p. 57). Metz aptly elucidates how his version of Afro-communitarian ethics of *Ubuntu* obliges "medical professionals such as doctors, nurses, bioethicists and public health researchers" to react in certain ways that will meaningfully address the health challenge of global warming in Africa and beyond (2016b). Be this as it may, Metz's theory is fundamentally anthropocentric in nature. A similar objection concerning Metz's theory concerning animal rights has been raised by Horsthemke (2015). And the response of Metz (2015) to the objection rightly explains that his theory does not ascribe equal moral status to humans and non-human animals and indeed other entities. Thus, I think that Metz's response to Horsthemke did admit that his theory is anthropocentric in a sense. This is not strange. For as Kelvin Behrens (2010) observes, there is a "pervasive presumption of anthropocentrism in African thought" (p. 465). Besides, the major concept, *Ubuntu*, which informs Metz's theory, is pervasively anthropocentric.

However, while Horsthemke (2015) quarrels with Metz's allocation of unequal moral status to humans and non-human animals, I think otherwise. My contention is that making *Ubuntu* the theoretical foundation of relational ethics tends to restrict the scope of relational ethics to purely human relations. Specifically, the cardinal principle of the theory explicitly limits the said harmonious relationship to human beings without an explicit position on the relationship between humans and other entities in the universe, including non-human animals and inanimate entities. This appears to me as a basic limitation that needs to be erased in an environment-friendly Africa-inspired ethical theory. Relational ethics stands in need of a comprehensive philosophical principle that encompasses both human-human relations and human-nature relations.

Explicating the common-links and prospects of the theories towards the challenge of GW

A careful study of the discussion so far will reveal that the common link between Asouzu's complementary reflection, Chimakonam's theories of African science and conversational thinking and Metz's relational ethics is their *emphasis on harmony*. First, the basic principles of the respective theories suggest in one way or the other that the disruption of any specific harmonious relationship is a wrong act that often produces negative results. Second, they agree to a large extent that the inherent human capacity to think or reason, equips humans with the ability to explore and discover the inherent relationship among all existents. Humans must identify and relate with all entities in the universe in specific ways to enable them to co-exist peacefully together for their common good.

Beyond their common emphasis on harmony, the primary concern of the theories is different from each other. Asouzu´s principle of integration deals primarily with the question of being, the nature of reality. The main interest of Chimakonam's theories of African science and conversational thinking are fundamentally on the question of scientific and philosophical methodology in the African place. Metz's relational ethics focuses primarily on the relationship between human beings. None of them taken alone can serve as a sophisticated Africa-inspired intellectual resource for understanding and reversing the tide of global warming today. For the challenge of GW is ontological, scientific, methodological, as well as ethical. My submission here is that one's conception of the nature of reality will to a large extent influence her/his methodology and relationship with reality. Modern science and technology (MST) and its methodological dispositions are the product of a given Western conception of the nature of reality and the ethics that emerged from it.

Accordingly, I endorse the view that, beneath the underexplored intellectual heritage of Africa is an approach to reality that can complement "the extremely narrow attitude implicit in the strict scientific approach characteristic of the West" (Agazzi, 1994, p. 13). A creative synthesis of the positive insights in the principles and theories examined could lead to the emergence of a novel philosophy of the environment and sustainable development from the African place. The said synthesis is possible because the positions of the respective theories complement and fill some of the lacunas inherent in one another. They all suggest that the quest for a good environmental ethics calls for the articulation and adoption of relevant philosophical principles, logical/methodological disposition and ethics that are sensitive to the necessary complementary interdependence of all existents in the universe. Besides the positions of Asouzu, Chimakonam and Metz discussed above, recent works in African environmental philosophy/ethics – notably, Kelbessa (2010, 2014, 2015) and Behrens (2010, 2011, 2012, 2014) – strongly support this view. The said creative synthesis is outside the scope of this work.

Conclusion

The basic contention of this work is that the present tide of GW is an ontological boomerang effect of the careless abuse of the inherent harmony among all existents in the universe by humans. I explain that the philosophical foundation and methodological disposition of MST (modern science and technology) influences GW. Given that MST operates from the prevalent Western philosophical standpoints, I argue for the development of an African philosophical framework that would serve as an alternative foundation and methodological disposition for EST (environment friendly science and technology). Along this line, I engage in a critical and creative conversation with selected works by three contemporary African philosophers, Asouzu, Chimakonam, and Metz, in search of relevant ideas. Thus, I identify certain ideas that can serve as a foundational resource for constructing an Africa-inspired philosophical framework that will help humans to understand and exploit nature to meet their needs without incurring ontological boomerang effects such as GW (severe global warming). My submission is that the conscious development and adoption of the African philosophical framework will help to repel the tide of GW in our time.

Ultimately, I maintain that the intellectual heritage of Africa houses lots of ideas that can be tapped to confront many of the challenges of humanity today. The continual underexploration of intellectual resources of Africa is detrimental to knowledge production. The study of African philosophy should therefore be globally encouraged.[7]

Notes

1 I borrow the term "Global Expansion of Thoughts" from Chimakonam (2015a, pp. 462–479).

2 The scientific consensus concerning the reality and major causes of the dangerous tide of global warming today is strongly buttressed by the recent publication of the findings of a UK-US team of leading climate scientists titled *Climate Change Evidence and Causes: An Overview from the Royal Society and the US National Academy of Sciences*.

3 See Chimakonam (2015b) and Agada (2015).
4 Note that by Western science, Chimakonam is referring to modern science.
5 The prevailing view in African philosophy is that the world is divided into three; the world of spiritual entities/deities, the world of human beings including other animate and inanimate but tangible existents, and the ancestral world of the living dead. And there is a necessary harmonious communal relationship between the inhabitants of these worlds.
6 Modelled after the likes of the Frankfurt school and the Vienna circle, the CSP is a philosophical movement inspired by the works of Panteleon Iroegbu, Campbell S. Momoh, and Innocent I. Asouzu but formally convened (at the University of Calabar, Calabar Nigeria) by Jonathan Chimakonam, as a professional body of (African and non-African) scholars who are committed to doing (or seek to do) African philosophy, and/or any aspect of African studies using the methodological dispositions and canons of Conversational philosophy (CP). For details, see Chimakonam (2014, 2015b, 2015c; Nweke, 2015a, 2015b) or visit www.unical.edu.ng/csp.
7 I would like to thank the conveners of the colloquium for the useful comments and insights that accompanied their acceptance of the proposal for this research. Many thanks to the colloquium panelists where the initial draft of this essay was presented especially Bruce Janz, Kelbessa Workineh, Ademola Fayemi, Lucky Ogbonnaya and Daniel Ekup-Nse Clement for their constructive comments, questions and suggestions. I also thank my colleagues at the Conversational School of Philosophy, Ada Agada, and Chukwueloka Uduagwu, for their useful observations and suggestions after reading the initial draft of this essay.

References

Acholonu, C. (1995). *The Earth Unchained*. Owerri: Afa Publications.

Agada, A. (2015). *Existence and Consolation: Reinventing Ontology, Gnosis, and Values in African Philosophy*. J. O. Chimakonam (Ed.). St. Paul, MN: Paragon House.

Agazzi, E. (1994). Philosophy, Humanity and the Environment. In H. O. Oruka (Ed.), *Philosophy, Humanity and Ecology Volume 1: Philosophy of Nature and Environmental Ethics* (pp. 3–14). Nairobi: ACTS Press and African Academy of Sciences.

Asouzu, I. (2004). *The Method and Principles Complementary Reflection in and Beyond African Philosophy*. Calabar: University of Calabar.

Asouzu, I. (2011). Ibuanyidanda and the Philosophy of Essence (Philosophy, the Science of Missing Links of Reality). *50th Inaugural Lecture of the University of Calabar*, University of Calabar Press, Calabar, May 18.

Behrens, K. (2010). Exploring African Holism With Respect to the Environment. *Environmental Values*, 19(4), 465–489.

Behrens, K. (2011). *African Philosophy, Thought and Practice and Their Contribution to Environmental Ethics*. Unpublished Doctoral Thesis, University of Johannesburg.

Behrens, K. (2012). Moral Obligations Towards Future Generations in African Thought. *Journal of Global Ethics*, 8(2–3), 179–191.

Behrens, K. (2014). An African Relational Environmentalism and Moral Considerability. *Environmental Ethics*, 36(1), 63–82.

Chimakonam, J. O. (2012). *Introducing African Science: Systematic and Philosophical Approach*. Bloomington, IN: Author House.

Chimakonam, J. O. (2014). Interrogatory Theory: Patterns of Social Deconstruction, Reconstruction and the Conversational Order in African Philosophy. *Filosofia Theoretica: Journal of African Philosophy, Culture and Religions*, 3(1), 1–25.

Chimakonam, J. O. (2015a). Transforming the African Philosophical Place Through Conversations: An Inquiry Into the Global Expansion of Thought (GET). *South African Journal of Philosophy*, 34(4), 462–479.

Chimakonam, J. O. (2015b). The Criteria Question in African Philosophy: Escape From the Horns of Jingoism and Afrocentrism. In J. O. Chimakonam (Ed.), *Atuolu Omalu: Some Unanswered Questions in Contemporary African Philosophy* (pp. 101–124). Lanham, MD: University Press of America.

Chimakonam, J. O. (2015c). Conversational Philosophy as a New School of Thought in African Philosophy: A Conversation With Bruce Janz on the Concept of 'Philosophical Space'. *Confluence: Journal of World Philosophies*, 3, 9–40.

DeGeorge, R. T. (1994). Modern Science, Environmental Ethics and the Anthropocentric Predicament. In H. O. Oruka (Ed.), *Philosophy, Humanity and Ecology* (Volume 1: Philosophy of Nature and Environmental Ethics, pp. 15–29). Nairobi: ACTS Press and African Academy of Sciences.

Edet, M. I. (2011). Being as Missing-Links. *Journal of Complementary Reflections: Studies in Asouzu*, 1(1), 28–32.

Horsthemke, K. (2015). *Animals and African Ethics*. New York: Palgrave Macmillan.

Jones, W., and Metz, T. (2015). The Politics of Philosophy in Africa: A Conversation. *South African Journal of Philosophy*, 34(4), 538–550.

Kebelssa, W. (2010). *Indigenous and Modern Environmental Ethics: A Study of the Indigenous Oromo Environmental Ethic and Modern Issues of Environment and Development*. Washington, DC: The Council for Research in Values and Philosophy.

Kebelssa, W. (2014). Can African Environmental Ethics Contribute to Environmental Policy in Africa? *Environmental Ethics*, 36(1), 31–61.

Kebelssa, W. (2015). Climate Ethics and Policy in Africa. *Thought and Practice: A Journal of the Philosophical Association of Kenya (PAK) New Series*, 7(2), 41–84.

Maslin, M. (2004). *Global Warming: A Very Short Introduction*. Oxford: Oxford University Press.

Metz, T. (2007). Towards an African Moral Theory. *Journal of Political Philosophy*, 15, 321–341.

Metz, T. (2015). How to Ground Animal Rights on African Values: A Reply to Horsethemke. A Presentation at the *International African Philosophy Conference* organized by the University of the Witwatersrand, South Africa and the University of Nairobi, Kenya on the theme: African Philosophy: Past, Present and Future. Senate Building, University of the Witwatersrand, Johannesburg, South Africa, September 9–11.

Metz, T. (2016a). An African Theory of Social Justice: Relationship as the Ground of Rights, Resources and Recognition. In C. Boisen and M. Murray (Eds.), *Distributive Justice Debates in Political and Social Thoughts: Perspectives on Finding a Fair Share* (pp. 171–190). London: Routledge.

Metz, T. (2016b). Climate Change in Africa and the Middle East in the Light of Health and Salient Regional Values. *Bioethical Insights into Values and Policy*, 4, 115–125.

Nweke, V. C. (2015a). Complementary Reflection vs. Binary Complementarity: An Interrogatory Discourse of the Problem of Anachronism in African Philosophy. Presented at the *International African Philosophy Conference* organised by the University of the Witwatersrand, South Africa and the University of Nairobi, Kenya on the theme: African Philosophy: Past, Present and Future. Senate Building, University of the Witwatersrand, Johannesburg, South Africa, September 9–11.

Nweke, V. C. (2015b). David A. Oyedola and the Imperative to Disambiguate the Term 'African Philosopher': A Conversation From the Standpoint of the Conversational

School of Philosophy. *Filosofia Theoretica: Journal of African Philosophy, Culture and Religions*, 4(2), 94–99.

Ramose, M. B. (2007). But Hans Kelsen Was Not Born in Africa: A Reply to Thaddeus Metz. *South African Journal of Philosophy*, 26(4), 347–355.

Reay, D. (2006). *Climate Change Begins at Home*. London: Macmillan.

Tschaepe, M. (2013). A Humanist Ethic of Ubuntu: Understanding Moral Obligation and Community. *Essays in the Philosophy of Humanism*, 21(2), 47–61.

12 Laudato Si and the ecological crisis

Ralph Okechukwu Madu

Introduction

Today, ecological crisis is proving to be one of the most daunting challenges, and perhaps the most arduous one that humans have faced since the beginning of life. The crisis is about our planet earth, where the survival of the biotic community, including human life, is increasingly under threat, caused by poor appreciation of the interrelatedness of all existents as well as their interdependence.

Ecology is a word frequently found on the lips of politicians, eco-warriors, business people, scholars, conservationists and even broadcasters. Yet trying to pin down exactly what they all mean is difficult. Thus, there is need for a concise explanation of the term. Etymologically, the word is derived from the Greek *oikos*, meaning "House," and in this original meaning the word would stand for household. Thus, ecology is the science which studies the structure and function of nature.

Later, ecology was seen narrowly as a sub-heading of biology that studied the interrelationship of living beings among themselves and their environment. In short, it is concerned with the preservation of certain endangered species or the creation of nature reserves, favourable conditions for the various ecosystems, the issue of biodiversity (Spicer, 2006, pp. 1–2). So understood, environment had meant for many people something outside and external to themselves and thus remained a peripheral and secondary concern (Kureethadam, 2014, p. 1). With such a mentality, humans fabricated machines and invented new tools for dominating the world and plundering its riches, which were seen as mere "natural resources" deserving of no respect.

But as awareness of the unwanted effects of industrial development grew, ecological discourse assumed a more encompassing dimension. It became increasingly obvious that it is no longer only the species but the earth as a whole that is ill and must be treated and healed. The interconnectivity between all living organisms within one household, and their relationship with the entire planet became more obvious. Everything is connected to everything.

Thus, more knowledge of the fact that the earth is our shared home increased. "All is related to all, at all points and at all times" (Boff, 1997, p. 106). This does not negate the fact that humans are the most "complex and unique expressions of the earth" (Boff, 1997, p. 106). "Man and woman are on the earth – thinking, hoping, loving, dreaming and entering into a phase in which decision is no longer by instinct but conscious" (Jantsch, 1980). This new vision of the earth, through one of its species, the human, opens a new phase of the evolutionary process, lays the foundation for a new ecological age, a new paradigm in which humanity initiates a new covenant of respect and kinship with other earth inhabitants: And as Berry (1988) notes, "this is the ultimate daring venture for the earth, this confiding its destiny to human decision, the bestowal upon the human community of the power of life and death over its basic life systems."

With this awareness, one would normally expect a radical change in the way humanity operates – political, economic, social, technological, life patterns, etc. One would have expected a radical change in the present course of economic development or drastic personal and cultural revolution, community life-style changes (Francis, 2015 [PF]). But that is not generally the case as the neoliberal economy, which is one of the major drivers of the mindless plundering of the planet through its paradigm of infinite growth, increasing production of material good, and the voracious consumption of them goes unchecked.

Even in the wake of the influx of environmental activism and many authoritative warnings from the scientific community, the major world summits on the crisis have always concluded with hollow promises, (Kureethadam, 2014, p. 6) (the First Earth summit in Rio de Janeiro 1992, Second Earth summit in Johannesburg in 2002 and the Third in Rio in 2012) (PF, 2015).

The paradoxical quagmire we find ourselves in today – the explosion of environmental activism and yet the constant information from the scientific world that the natural processes that sustain life on earth are at the verge of collapse – calls for urgency in the handling of the ecological crisis. It calls for both individual and global conversion and reorientation to our attitude towards nature.

The thrust of this chapter, therefore, is about the crisis facing our common home, about how the capacity of this to be truly a home to humanity and other creatures is being jeopardised. A major thrust in the debate would be the unravelling of the matter by Pope Francis in his recent Encyclical, *Laudato Si*, which we shall elaborate on, and make references to, across all the sections of this work (PF, 2015).

But in order to give a fair treatment to the matter, we shall dedicate a section of the chapter to trying to get a proper understanding of this deeply existential crisis. "The fact that the ecological crisis is only getting worse even after decades of study, diagnosis and mitigation efforts means that we have not sufficiently understood the malaise in the first place" (Kureethadam, 2014, pp. 7–8). Since the ecological crisis is not merely a physical problem but has a human face, the morality of the human lifestyle deserves some consideration along the lines of *Laudato Si*. As this shall necessarily raise the issue of eco-justice and human rights, this chapter shall discuss, in a rather cursory manner, the impact of the ecological crisis on the poor, women, children, the aged, the infirm, minorities, developing states, future generations etc., as well as "the structural forms of poverty that exist throughout the world" (Statement of Holy See Delegation to the UN Conference,

June, 4, 1992 [SHSDU]). Finally, the effects of the eco-crisis on the poor and struggling people of Africa, especially Nigeria, shall be highlighted.

Crossing the thresholds

As we noted earlier, there is a need to properly articulate the ecological problem so that we can appreciate the enormity of the matter and be better positioned to proffer effective solutions. Pope Leo XIII (1891), noted that "nothing is more useful than to look upon the world as it really is" (no. 18). A detailed study of this aspect of our discourse would necessarily cover a long historical period, ranging from the Agricultural to the Industrial ages and culminating in the space-travel epoch of the human development processes. But we shall make do with a rather perfunctory approach.

In the earliest times when agriculture was central, the early civilisation era, for thousands of years, humanity had "existed in the sweetest of sweet spots. The temperature has barely budged, globally averaged, its swing in the narrowest of range, and between fifty-eight and sixty degrees Fahrenheit . . . it was the "correct" temperature for the marvellously diverse planets that seems right to us" (Mckibben, 2010, p. 1). Development in Agriculture provided a greater abundance of food and sustained man's primordial relationship with the cosmos. Using Neolithic technology, man was able to "domesticate plants and animals, breeding them to render them more productive. People manipulated once-world plants and began to raise wheat, rice, corn, barley, millet, peas, beans, and other familiar farm animals. Grasslands were ploughed and rivers were diverted for irrigation" (Kureethadam, 2014, pp. 42–43). Around this time, a wide variety of languages, social, political and religious systems developed, including a wide range of sedentary communities.

The industrial revolution beginning from the 15th–18th centuries gave rise to a radical change in humanity's relationship with the physical world. The birth of modern science and technology brought improvements to human life and provided humanity the technology to harness the resources of nature in an unprecedented way that left heavy imprints on the Earth and its natural processes.

The 19th and the 20th centuries marked the period of great acceleration and globalisation, a social, economic and political process that further disrupted the natural equilibrium that nature seemed to have been managing. With the possible exception of nuclear weapons, capitalism is the most powerful of human inventions. As an expression of the restless soul of European civilisation and as the urge to discover, create and conquer, capitalism has now spread to every part of the world, breaking down all barriers standing on its way. Thus, the agents and forces of the system, the mega corporations, the International Monetary Fund (IMF), the World Bank (WB) and the General Agreement on Tariff and Trade (GATT), all of them off-shoots of the 1944 Bretton Woods Agreement, including the Geneva 1993 established World Trade Organisation (WTO), have aggressively integrated all economies of the world into a single capitalist system, exposing the global order to gross mutual vulnerability. It is this process that is generally termed globalisation (Madu, 2015, p. 164). Other globalising phenomena include the Trans-Atlantic slave trade, the imperial expansion of Europe, the two world wars and the cold war.

Under the banners of neo-liberalism, the proportion of those in distress "increases, so does the probability of human insecurity and vulnerability." As Alain (1996) uncomfortably noted, "All things are now being transformed into currency. What enters the system as a living thing comes out as a commodity, a dead product" (p. 119).

As Kureethadam observes, "The ecological crisis has many facets that are interlinked but still distinct" (Kureethadam, 2014, p. 81). Climate change is and remains the most challenging and indeed an entry point into the whole of the ecological crisis (Kureethadam, 2014, p. 85). Geological, archaeological, oral and written histories all attest to the dramatic challenges posed to the past societies from unanticipated changes in temperature, precipitation, winds and other climatic variables (Haffner, 2008, p. 49). "Global warming" is a specific example of climate change, and studies have shown that most of the observed increase in globally averaged temperatures since the mid-20th century is very likely due to the observed increase in anthropogenic greenhouse gas concentrations via the greenhouse effects (Haffner, 2008, p. 45).

As the study by the Royal Society, a scientific academy, clearly attests, "Changes in climate" have significant implications for present lives, for future generations and for ecosystems on which humanity depends (2010, September 30, para 1). There is the general impression that the threat to the earth as a result of climate change manifests itself gradually and can only be full blown in the distant future. However, this is not correct because scientists have warned forthrightly that "society may be hulled into a false sense of security by smooth projections of global change. Our synthesis of present knowledge suggests that a variety of tipping elements could reach their critical point within this century under anthropogenic climate change" (Lenton et al., 2007, pp. 1789–1792).

Desertification is yet another facet of the ecological crisis. Even though closely linked to climate change, it has other causes, such as deforestation and land degradation on account of intensive agriculture. It is also one of the causes of the extinction of species. As Pope John Paul II (1990, January 1) succinctly observed, "delicate ecological balances are upset by the uncontrolled destruction of animal and plant life or by a reckless exploitation of natural resources. It should be pointed out that all of this, even if carried out in the name of progress and well-being, is ultimately to mankind's disadvantage" (no. 7). The recent Millennium Ecosystem Assessment 2005 reveals that about 60% of the ecosystem services that support life are being degraded or used unsustainably (Spicer, 2006, p. 89). Through agriculture, logging, industry, mining, pasture, building roads, settlements, etc., humans have persistently restructured nature and thus contributed to animal extinction. The Earth is generous and pays back abundantly when it is handled in keeping with its own inner logic (Boff, 1997, p. 134).

Pollution of land, water and air is another factor. Again, humanity is consuming natural resources faster than ecosystems can regenerate them. This is the result of an unnatural and reductionist vision which at times leads to a genuine contempt of man (John Paul II, 1990). Pollution is a modern occurrence which began with the industrial revolution and has attained great heights within the last decade of economic expansion. The "environment is devastated and development thwarted by the outbreak of wars, when internal conflicts destroy homes, fields and factories, when intolerable circumstances force millions of people to desperately seek refuge away from their lands, when minorities are oppressed, and the most vulnerable are neglected or abused" (SHSDU). The exposition of these concerns by Pope Francis in his Encyclical letter *Laudato Si* will be our preoccupation in the next sections.

Laudato Si and the Church's social and environmental agenda

Over the years, the Catholic Church has constantly exercised grave concern over the poor, the environment and the collective well-being of humanity. This has constituted the major thrust of her social teaching, encapsulated in numerous papal encyclicals and pastoral documents. Such doctrine not only helps governments and other prevailing agencies change their overbearing attitudes, but also helps Christians assume their responsibilities towards the common good. Catholic social teaching is, therefore, not a static or fixed doctrine, but a dynamic application of Christ's teaching to the changing realities and circumstances of human actualities and cultures. (Abue, 2005, p. 126) The Encyclical *Laudato Si* by Pope Francis is one of such Church's social documents.

The Encyclical takes its name, *Laudato Si*, from the invocation of Saint Francis of Assisi, "Praise be to you, my Lord" which in the *Canticle of the Creatures* reminds us that the earth, our common home "is like a sister with whom we share our life and a beautiful mother who opens her arms to embrace us." "This sister now cries out to us because of the harm we have inflicted on her by our irresponsible use and abuse of the goods with which God has endowed her." Her cry, united with that of the poor, stirs our conscience to "acknowledge our sins against creation" (PF, no. 8). The appropriate response to this cry, according to Pope Francis, is what St John Paul II already called a "global ecological conversion" (PF, no. 5). In this, St Francis of Assisi is "the example par excellence of care for the vulnerable and of an integral ecology lived out joyfully and authentically. [. . .] He shows us just how inseparable is the bond between concern for nature, justice for the poor, commitment to society, and interior peace" (PF, no.10).

Laudato Si is constructed on the presupposition that creation is essentially good and thus has a positive value within the Christian economy of salvation and is thus developed around the concept of integral ecology, as a paradigm able to articulate the fundamental relationships of the person: with God, with one's self, with other human beings, with creation. It is a six-chapter document. As Pope Francis (2015) explains, the movement towards ecological reorientation and conversion starts by listening spiritually to the results of the best scientific research on environmental matters available today, "letting them touch us deeply and provide a concrete foundation for the ethical and spiritual itinerary that follows" (PF, no. 15). Science is a useful tool by which we can listen to the cry of the earth.

In Chapter 2, he picks up the wealth of the Judeo-Christian tradition, particularly in biblical texts and in theological reflection on them. The analysis is then directed in chapter III towards "the roots of the present situation, so as to consider not only its symptoms but also its deepest causes." The objective is to develop an integral ecological outline (ch. IV) which, in its various dimensions, includes "our unique place as human beings in this world and our relationship to our surroundings." Insisting on clear and inclusive dialogue, Pope Francis proposes in chapter V a series of guidelines for the renewal of international, national and local policies, for better decision-making processes in the public and business sector, for the relationship between politics and economy and that between religion and science.

Finally, based on the conviction that "change is impossible without motivation and a process of education," chapter VI proposes "some inspired guidelines for human development to be found in the treasure of Christian spiritual experience." Along this line, the Encyclical offers two prayers, the first to be shared with believers of other religions and the second among Christians. The Encyclical concludes, as it opened, in a spirit of prayerful contemplation.

Despite practical relativism and the modern consumer culture, the Pope is still hopeful: "All is not lost. Human beings, while capable of the worst, are also capable of rising above themselves, choosing again what is good, and making a new start, despite their mental and social conditioning. . . . No system can completely suppress our openness to what is good, true and beautiful, or our God-given ability to respond to his grace at work deep in our hearts. I appeal to everyone throughout the world not to forget this dignity which is ours" (PF, no. 205). Changes in lifestyle and consumer choices can bring much "pressure to bear on those who wield political, economic and social power" (PF, no. 206). "If we can overcome individualism, we will truly be able to develop an alternative lifestyle and bring about significant changes in society" (PF, no. 208).

Each of the six chapters of *Laudato Si* addresses a particular topic using a specific method, but throughout the entire Encyclical, some principal concepts are continually taken up and enriched:

- the intimate relationship between the poor and the fragility of the planet,
- the conviction that everything in the world is connected,
- the critique of new paradigms and forms of power derived from technology,
- the call to seek other ways of understanding the economy and progress,
- the value proper to each creature,
- the human meaning of ecology,
- the need for forthright and honest debate,
- the serious responsibility of international and local policy,
- the throwaway culture and the proposal of a new lifestyle.

The dialogue that Pope Francis proposes as a method for addressing and resolving environmental problems is followed throughout the Encyclical. It refers to contributions by philosophers and theologians, not only Catholic but also the Orthodox Patriarch Bartholomew and Protestant French thinkers such as Paul Ricoeur, as well as the Islamic mystic Ali Al-Khawas. The same occurs in the key of that collegiality that Pope Francis has proposed to the Church since the beginning of his papacy: alongside the references to the teachings of his predecessors and of other Vatican documents there are many ideas taken from numerous Episcopal Conferences from all continents (Ivereigh, 2015, June 18).

At the heart of *Laudato Si*, we find this question: "What kind of world do we want to leave to those who come after us, to children who are now growing up?" (PF, no. 160). Pope Francis continues: "This question does not have to do with the environment alone and in isolation; the issue cannot be approached piecemeal." This leads us to ask ourselves about the meaning of existence and its values at the base of social life: What is the purpose of our life in this world? Why are we here? What is the goal of our work and all our efforts? What need does the earth have of us? According to Pope Francis, "Unless we struggle with these deeper issues I do not believe that our concern for ecology will produce significant results."(PF, no. 160).

The Encyclical is no less apprehensive of international dynamics: "Recent World Summits on the environment have failed to live up to expectations because, due to lack of political will, they were unable to reach truly meaningful and effective global agreements on the environment" (no. 166). Instead, as popes have repeated several times starting with *Pacem in terris* (April 11, 1963), what is needed are forms and instruments for global governance (no. 175): "an agreement on systems of governance for the whole range of the so-called global commons'" (no. 174).

It is clear that, after *Laudato Si*, the examination of conscience should include a new dimension. One should regularly review, not only how one has lived in communion with God, with others and with oneself, but also with all creatures and with nature.

The religious and moral nexus of *Laudato Si*

The attention that the media gave to the Encyclical before its publication primarily concentrated on aspects tied to the environmental policies currently being discussed on the global agenda. This led to a series of criticisms that preceded its publication. One of such reactions was from the former Florida Governor Jeb Bush. The *Washington Times* quoted him as saying: "I hope I'm not going to get castigated for saying this by my priest back home, but I don't get economic policy from my bishops or my cardinals or my pope. . . . And I'd like to see what he says as it relates to climate change and how that connects to these broader, deeper issues before I pass judgment" (June 17, 2015). Also, the former Pennsylvania Senator, Rick Santorum, a 2016 presidential aspirant, was reported as advising Pope Francis to "leave science to the scientists." The above two politicians questioned the competence of a Pope who has a Master's degree in Chemistry to speak on scientific matters. They questioned the role of religion and the church in the formulation of public policy.

Certainly *Laudato Si* can and must have an impact on important and urgent decisions to be made in this area. However, the magisterial, pastoral and spiritual dimensions of the document must not be put in second place. Its value, breadth and depth cannot be reduced to the mere scope of determining environmental policies. As Pope Francis pointed out, "there needs to be a distinctive way of looking at things, a way of thinking, politics, an educational programme, a lifestyle and a spirituality which together generate resistance to the assault of the technological paradigm" (PF, no. 111). The richness and depth of *Laudato Si* derives largely from being rooted in the rich heritage of Scripture and Tradition.

The banquet motifs in the Scriptures (Isaiah 22:13; 25:6, Matt 26:29, Lk:14, Acts 10:41) call to mind all that is most distinctly human, gracious, festive and fun. Only humans make and enjoy wine. Only humans laugh.

We have the phrase "the banquet of life," and we mean by it the vastness, the abundance, the diversity, the harmony, and the astounding fittingness and balance of the universe we live in. It was on the sixth day of Creation that God called humanity to the banquet of life. Everything was ready, the table, so to speak, laid, when man was at last invited, the image and likeness of the very God who made the feast. The human was to be the first, but was rather invited last of all.

The scientific study of human origins is called paleoanthropology. Paleoanthropology tells us that we modern humans, the species Homo sapiens arrived on the scene in what is today Africa only about 100,000 years ago. When you consider that almost anywhere in this world of ours you can find a rock to sit on that is several billions of years old, the age of our race is a mere eye blink old. One hundred thousand years is not much, and it was more recently that we began to actualise in our behaviour the potency that was ours as humans. It was only maybe 50,000 years ago that humans began to behave as the Image of God, that the human arrived at the banquet with the commission to serve the rest.

Human beings are the walking representations of God, with a call to responsible care and stewardship to the world and to the other species. The glory of creation and its perfection depend on our success as humans. Our libraries, as well as the surface of our earth and the quality of our waterways and of the air we breathe provide so many records of how we've done. They are so many mirrors in which we can see on our faces either honour or shame.

For many of us, rather than the Bible, it is bumper stickers that are a source of wisdom. One I saw the other day says so much in ecological relations: "Live simply, so that others may simply live."

In his book *The Creation*, Wilson (2006) calls religion and science "the two most powerful forces in the world today." According to him, "If religion and science could be united on the common ground of biological conservation, the problem would soon be solved. If there is any moral precept shared by people of all beliefs, it is that we owe ourselves and future generations a beautiful, rich and healthful environment" (p. 5). Religious leaders help shape awareness of and gratitude for the complex and tender sphere (Wilson, 2006, p. 27).

Thus, the ecological crisis is not simply economic and technological, it is moral and spiritual (Haffner, 2008, p. 287). Pope John Paul II (1990) maintained that the environmental problem is a moral problem (no. 6–7, 15). Moral accountability is necessary because what makes society possible is the fact that there are commonly recognised fundamental norms that establish limits to actions, which in principle, must not be transgressed, even if they would lead to certain advantage (Obi, 2005, pp. 49–116).

This is against the prevailing tendency to believe that every advance in technology and power is good for society, which would be true only if there were also an advance in human responsibility, values and conscience. The empirical sciences do not completely explain life, and technical solutions are ineffective "if we lose sight of the great motivations which make it possible for us to live in harmony, to make sacrifices and to treat others well" (PF, no. 200). These values are often expressed in the language of religions. That notwithstanding, religions must enter into "dialogue among themselves for the sake of protecting nature, defending the poor, and building networks of respect and fraternity" (PF, no. 201). At the same time, dialogue among the sciences helps to overcome disciplinary isolation. "An open and respectful dialogue is also needed between the various ecological movements" (PF, no. 201), the Pope maintains. Dialogue requires patience, self-discipline and generosity.

Sin exists in all realms of human existence, and therefore also in the relationship between human beings and their environment. (Haffner, 2008, p. 244). According to Patriarch Batholomew (1997, November 8) "For humans . . . to destroy the biological diversity of God's creation . . . for human beings to degrade the integrity of the Earth by causing changes in its climate, by stripping the Earth of its natural forests or destroying its wetlands . . . for humans to injure other humans with disease . . . for humans to contaminate the Earth's waters, its land, its air, and its life with poisonous substances . . . these are sins." For "to commit a crime against the natural world is a sin against ourselves and a sin against God" (Speech at Santa Barbara Greek Orthodox Church). Thus, while Pope Francis does not pretend to proffer technical solution to the crisis, it behooves the Church to offer spiritual, intellectual and moral guidance. That, precisely, is what *Laudato Si* and many such encyclicals in the long tradition of Catholic social teaching set out to achieve.

We may search for a technological answer to the multiple ecological problems we face, but the questions are really human ones (Tull, 2013, p. 3). As Pope Francis strongly notes: "The violence present in our hearts, wounded by sin, is also reflected in the symptoms of sickness evident in the soil, in the water, in the air and in all forms of life. This is why the earth herself, burdened and laid waste, is among the most abandoned and maltreated of our poor; "she groans in travail" (PF, no. 2).

Religious authority has often seen the danger inherent in "progress." The forces that have progressively and imprudently, at different times and places, broken down the bonds that hold nature together have been easily labelled expansion of commerce, industrialisation, globalisation etc., and these are often justified as implying progress, growth and inevitability. But *Laudato Si* questions the senseless "rapidification," noting that "although change is part of the working of complex systems, the speed with which human activity has developed contrasts with the naturally slow pace of biological evolution. Moreover, the goals of this rapid and constant change are not necessarily geared to the common good or to integral and sustainable human development. Change is something desirable, yet it becomes a source of anxiety when it causes harm to the world and to the quality of life of much of humanity" (PF, no. 18).

In all these, there is a feeling of a grave loss of faith, temperance and the loss of an ultimate goal (PF, no. 77–83). The modern scientific person often does not accept that there is in human nature a level of reaching into a higher dimension of meaning where incarnate understanding and expression are inadequate. Consequently, not everyone would be happy with a response that uses religious language, as is evident in *Laudato Si*. "The extinction of faith is being recognised as the supreme danger to the spiritual health and even to the material existence of the Western body social – a deadlier danger by far than any of our hotly canvassed and loudly advertised political and economic maladies. To refill the spiritual vacuum is the most formidable and the most urgent of the problems that are remorselessly crowding upon the present generation" (PF, no. 216–220).

Today, the 'absolutisation' of what works has not only impinged on human conscience and morality but has also given vent to serious neglect of the ethical implications of man's activities. Pope Francis cautions: "Many things have to change course, but it is we human beings above all who need to change. We lack an awareness of our common origin, of our mutual belonging, and of a future to be shared with everyone. This basic awareness would enable the development of new convictions, attitudes and forms of life. A great cultural, spiritual and educational challenge stands before us, and it will demand that we set out on the long path of renewal" (PF, no. 202).

As we have noted earlier in this chapter, the ecological crisis we face is but one small sign of the larger spiritual crisis today. We can heal the wounds in nature only if we also at the same time heal all human relationships (PF, no. 120). No wonder, then, that Pope Francis enjoins self-restraint (PF, no.165), cultural revolution (PF, no. 114) protection of human life (no. 120), the value of labour (PF, no. 124), the need to invest in people (PF, no. 128), restraint in excessive possession of financial power (PF, no.129), respect for the most disadvantaged population (PF, no.182–188), change in lifestyle and consumer choices (PF, no. 203), ecological conversion (PF, no. 216–221), break with the logic of violence, exploitation and selfishness (PF, no. 230), love for society and commitment to the common good (PF, no. 231) as important ethical-moral principles that constitute the integral ecology which he enunciates. These, no doubt, call for in-depth dialogue, intellectual dialogue between science and religion, between society and the Church, because the ethical principles we need may very well come from the religious sphere (PF, no. 200).

The disparity in distribution of eco-damages

We have noted in this chapter the moral character of the ecological crisis, the fact that eco-crisis does not only affect the physical fabrics of the planet but also jettisons the social and ethical bonds that unite humanity. Thus, any true attempt at solving the ecological crisis must be distinctively moral in vision and approach. It must take cognizance of the intricate issues of justice, equity and solidarity. This, by extension, should address the issues that guarantee human dignity: the rights of the poor and developing countries.

There exists a worrying disproportionality in the way ecological crisis affects peoples. While the rich minority is the major cause of the crisis, the vast majority of the poor and vulnerable are the unjust victims. According to the World Population Report, the world's richest half billion people – about 7% of the global population – are responsible for about 50% of the world's carbon dioxide emissions. Meanwhile the poorest 50% are responsible for just 7% of the emissions (UNFPA, State of World Population, 2011, p. 94).

Disparity in the distribution of ecological damages between the rich and the poor nations is further vividly enunciated by Caritas Internationalis: "The effects of climate change are already a reality for many people, particularly for the world's poorest and most vulnerable" (*Climate Change: Seeking a Global Ethics*, 2009, p. 6).

As Pope Francis notes, "The deterioration of the environment and of the society affect the most vulnerable people on the planet" (PF, no. 48), the greater part of the world's population. In international economic-political debates, they are considered "merely as collateral damage" (PF, no. 49).

As Hessel notes, "The 1987 UCC's study on environmental racism clearly shows that there exists a functional relationship between poverty, illiteracy, apartheid, racism, powerlessness, ineffectiveness and chemical industry's assault on the environment, communities, countries that are poorer, less informed, less aware, less organised and less politically influential become likely targets for the dumping of toxic wastes and other abuse from pollutes" (Hessel, 1996, p. 166). There is therefore a vicious cycle created between poverty and environmental degradation, one feeding on the other.

Studies have also shown the existence of a perverse logic at work in the impoverishment that affects most of the world's population (Boff, 1997, p. 101), exploitation of workers and pillaging of whole nations, a logic that tends to expand and deepen the abyss between the North and the South, the rich and the poor.

Boff (1997) argues that "it no longer suffices merely to adjust technologies or to reform society while keeping the same basic logic." There has been huge progress, but it is profoundly "inhuman," he maintained. Recent data reveal the stark inequalities inherent in the current world order, that 20% of the earth's population consumes 80% of the world supplies, while the other 80% have only 20%.

This "consumeristic selfishness" and devastation of the poor may meet some resistance, a rebellion, destruction, struggle for survival that might threaten everyone and destroy everything unless there is a collective decision to change the course of things and mitigate the logic of "exclusionary accumulation," the diversity between the geographical and ideological poles of our world and to work for more solidarity and common good of all nature (PF, no. 50).

The overconsumption patterns of a small minority of the world's population endanger our common home. "The poor live in the most polluted places, in proximity to industrial areas, close to transport lines, in neighbourhoods poorly serviced on water supply or garbage collection" (Kempf, 2008, p. 41). According to Diamond (2008), "The estimated 1 billion people who live in developed countries have a relative per capita consumption rate of 32. Most of the world's other 5.5 billion people constitute the developing world, with relative per capita consumption rates below 32, mostly down toward 1" (*The New York Times*, 2008, January 2).

The ecological segregation between the rich and the poor nations is further buttressed even in areas of intervention grants. "For example International funding for adaptation in developing nations in 2008 amounted to US$26 million roughly the same amount UK spends in flood defenses in a week" (UNDP Human Development Report, 2007/2008, 25). That is a great ethical tragedy that cries for global and integral attention and calls for the rich nations to assume greater responsibility in the efforts to heal our planet. "Basic justice and fairness imply that those who amassed wealth and other benefits through the nearly unrestricted extraction and use of petroleum, coal, and chemicals bear ethical responsibility for addressing the harms produced by the industrial greenhouse effect" (Cuomo, 2011, pp. 697–698).

Pope John XXIII condemned this increasing inequality, calling for more solidarity. According to him, it is "impossible for wealthy nations to look with indifference upon the hunger, misery and poverty of other nations whose citizens are unable to enjoy even elementary human rights . . . [It is] nothing less than an outrage of justice and humanity to destroy or to squander goods that other people need for their very lives."

The reference to poor nations that we have made so far includes the African continent. Asoanya (2004) reports that "The continent contains the world's largest expanse of dry lands, covering roughly two billion hectares of the continent or say 65% of Africa's total land area" (p. 113). According to a UNEP 1994 survey report, almost one-third of Africa's land is hyper-arid, and dry semi-humid areas, housing about 400 million people, two-thirds of the continent's total. Thus in Africa, poverty and environmental decline are like Siamese twins. The extent of these two maladies gave vent to Lloyd's fears that "environmental degradation will never be really checked in Africa while five million children die each year" (Lloyd, 1994, p. 23).

In effect, poverty can drive ecological deterioration when desperate people over-exploit their resource base, sacrificing the future to salvage the present, when Africa continues to attract massive transfer of "reject technologies" with high greenhouse gas emission. Even in circumstances where foreign multinationals are exploring in Africa, many of their operations and materials are outdated, in poor condition and would be illegal in other parts of the world (Bisong, 2005, p. 36).

The Nigerian scenario

The picture painted so far does not necessarily imply that Africa and its leaders are unperturbed about the extent of the danger, but rather that the level of the "crisis of sustainability and the level of corruption are overwhelming. The dilemma is worsened by poverty, insufficient education on environmental cleanliness, powerlessness, ineffectiveness, lack of proper political institutions, lack of will and savvy of social consciousness, external encroachments and a sense of alienation and anomie in the population (Obi, 2005, p. 115). For example, in his May 2016 presidential broadcast, President Mohammadu Buhari of Nigeria avowed, "we want to assure our neighbours, friends and development partners that Nigeria is firmly committed to democratic principles. We are ready partners in combating terrorism, cybercrimes, control of communicable diseases and protection of the environment. Following on the Paris Agreement COP 21, we are fully committed to halting and reversing desertification. Elsewhere, we will intensify efforts to tackle erosion, ocean surge, flooding and oil spillage which I referred to earlier by implementing the United Nations Environment Programme (UNEP) report." While this may convey the impression of a high-level commitment, the execution remains a matter of great debate. This is one in a long list of avowals by African leaders showing the awareness of the dangers inherent in the abuse of our planet. But mere theory without corresponding action would never produce any results.

Even in the face of the above avowal, the ecological disaster in the South regions of Nigeria resulting from oil spillage and gas-flaring, especially in the Niger Delta region, remains a worrisome scenario. The Niger Delta phenomenon is a typical example of what is happening in many parts of Nigeria and Africa, namely, our inability to build bridges to share one another's concern including that of our organic beings. Other realities of ecological degradation in Nigeria include

- Poor waste management; in towns, streets and even highways are littered with garbage at times to the extent of blocking the roads and making them smelly.
- Poor disposal of hardware, plastics and poor quality material of all genres that are used only once and thrown away.
- Unregulated manner of tree-logging, destroying rain forests, felling of trees for subsistence agriculture, timber and production of charcoal sold in bags everywhere along the roads. The traditional African idea of forest as sacred reserves is no longer obtainable even among businessmen and governments (Ake, 2001, p. 6).
- Poor power generation, which has given rise to the ubiquitous presence of generating sets ranging from the famed "*I big pass my neighbour*" to the 300–2000 KVA machines which pollute not only the environment with exhaust fumes but also saturate it with deafening noise.
- Wildlife and biodiversity depletion through deforestation and oil spillage.
- Disposal of industrial and toxic wastes in Nigeria by Western countries.
- Indiscriminate and senseless grazing.

What is obvious here is the inability of the entire nation to control the degradation of the ecosystem; thus, our boom has become our bane including diversities in culture and religion (Odemene, 2005, p. 153).

In the Niger Delta, multinational (Shell and others) high-pressure pipelines are constructed above ground through villages and crisscross over land that was once used for agricultural purposes, rendering it economically useless. Many pipelines pass within meters from people's homes, a blatant disregard for the people's rights and well-being. Unfortunately, statistical data also show that of the total number of oil spills recorded by Shell, a company that operates in more than 100 countries, 40% were in Nigeria. (Bisong, 2005, pp. 36–37). And as Obi (2005) very succinctly noted, "This aggressive and unregulated exploitation of natural resources in the name of development and economic growth is un-African, unethical and inconsonant with the reality of a finite world and our responsibility to be frugal in consumption, so that future generations too can effectively build their welfare on what remains of these resources when we are gone" (p. 52).

Conclusion

Our assessment of the ecological crises in the light of Pope Francis' Encyclical letter *Laudato Si* highlighted the fact that economic development and sound environmental management are but complementary aspects of the same agenda, which can be made meaningful only by an integral ecology. "An integral ecology is also made up of simple daily gestures which break with the logic of violence, exploitation and selfishness" (PF, no. 230). It is perhaps most pertinent at this point to re-echo by way of conclusion the appeal of Pope Francis on the way forward in the management of eco-crisis:

> We must regain the conviction that we need one another, that we have a shared responsibility for others and the world, and that being good and decent are worth it. We have had enough of immorality and the mockery of ethics, goodness, faith and honesty. It is time to acknowledge that light-hearted superficiality has done us no good. When the foundations of social life are corroded, what ensue are battles over conflicting interests, new forms of violence and brutality, and obstacles to the growth of a genuine culture of care for the environment. . . . In order to make society more human, more worthy of the human person, love in social life – political, economic and cultural – must be given renewed value, becoming the constant and highest norm for all activity. In this framework, along with the importance of little everyday gestures, social love moves us to devise larger strategies to halt environmental degradation and to encourage a "culture of care" which permeates all of society. When we feel that God is calling us to intervene with others in these social dynamics, we should realise that this too is part of our spirituality, which is an exercise of charity and, as such, matures and sanctifies us.
>
> (PF, no. 229–231)

References

Abue, P. (2005). *Oil and Gas Exploitation and Nigeria's Forest Resources in Nigeria's Forest Resources – Stakeholder Participation and Collaboration in Forest Management as Antidote*. In J. Wangbu (Ed.), *Niger Delta, Rich Region Poor People* (pp. 117–142). Enugu: Snaap Press.

Ake, C. (2001). *Democracy and Development in Africa*. Ibadan: Spectrum Book.

Alain, B. (1996). Confronting Globalization. Telos, June 20, Issue 108:117–137; doi:10.3817/0696108117

Asoanya, A. (2004). *The Ecological Crisis in Africa as Challenge: Lasting Cultural and Structural Development*. Frankfurt: Peter Lang.

Berry, T. (1988). *The Dreams of the Earth*. San Francisco, CA: Sierra Club.

Bisong, K. (2005). *Profit Maximization and Abuse of Human Rights in Oil Rich Niger-Delta of Nigeria*. In J. Wangbu (Ed.), *Niger Delta, Rich Region Poor People* (pp. 29–47). Enugu: Snaap Press.

Boff, L. (1997). *Cry of the Earth, Cry of the Poor*. New York: Orbis Books.

Caritas Internationalis. (2009). *Climate Change: Seeking a Global Ethics*, Retrieved from http://www.caritas.org/2010/06/climate-change-seeking-climate-justice/

Cuomo, C. J. (2011). Climate Change, Vulnerability, and Responsibility. *Hypatia*, 26, 697–698.

Diamond, J. (2008, January 2). What's Your Consumption Factor? *The New York Times*, 17.

Haffner, P. (2008). *Towards a Theology of the Environment*. Herefordshire: Gracewing Pub.

Hessel, D. T. (1996). Theology for Earth Community – a Field Guide. New York: Orbis Bks.

Ivereigh, A. (2015, June 18). *Laudato Si: A Summary of Pope Francis's Sweeping Eco-Encyclical*. www.catholicvoices.org.uk

Kempf, H. (2008). *How the Rich Are Destroying the Earth*. Devon: Green bks.

Kureethadam, J. (2014). *Creation in Crisis: Science, Ethics, Theology*. New York: Orbis Books.

Lenton, T., Held, H., Kriegler, E., Hall, J. W., Ramstorf, S., and Schellnhuber, H. J. (2008). Tipping Elements in the Earth's Climate System (1786–1793). *Proceedings of the Natural Academy of Sciences*. Vol. 105, no. 6. Pp. 1786-1793. doi: 10.1073/pnas.0705414105

Lloyd, T. (1994). *Africa in Crisis: The Causes, the Cures of Environmental Bankruptcy*. Nairobi: Kenya English Press.

Madu, R. (2015). *African Culture and Other Culture in the Face of Globalization*. In J. Fadugba-Pinheiro and E. Adeojo (Eds.), *A Dynamic Visioner* (pp. 164–181). Lagos: Gazub.

Mckibben, B. (2010). *Earth: Making a Life on a Tough New Planet*. New York: Times-Books/Henry Holt and Co.

Obi, E. (2005). *Environmental Justice and the Quest for an Equitable and Sustainable Economic World Order: Reassessing the Engagement of Multinational Corporations in the South-South of Nigeria*. In J. Wangbu (Ed.), *Niger Delta, Rich Region Poor People* (pp. 49–116). Enugu: Snaap Press.

Odemene, A. (2005). The Interrelatedness of Creation, Ecology and Culture: A Quest for a Sustainable Development in the Niger Delta. In J. Wangbu (Ed.), *Niger Delta, Rich Region Poor People* (pp. 143–155). Enugu: Snaap Press.

Patriarch Bartholomew. (1997, November 8). *Speech at Santa Barbara Greek Orthodox Church)*. http:/?www'patrarchate.org

Pope Francis. (2015). *Laudato Si: On Care for Our Common Home*. Vatican: Libreria Editrice Vaticana.

Pope John Paul II. (1990, January 1). *Peace With God the Creator, Peace With All of Creation*. World Day of Peace Message, 7.

Pope Leo XIII. (1891, May 15). *Rerum Novarum: Encyclical of Pope Leo XIII on Capital and Labor*. Retrieved from w2.vatican.va/content/leo-xiii/en/../hf_l-xiii_enc_15051891_rerum-novarum.html

Rio de Janeiro, The Royal Society. (2010, September 30). *Introduction, Climate Change: A Summary of the Science (para 1).*
Spicer, J. (2006). *Biodiversity: A Beginner's Guide*. Oxford: One World Pub.
Statement of Holy See. (1992, June 4). *Delegation to the UN Conference on Environment and Development.*
Tull, P. (2013). *Inhabiting EDEM: Christians, the Bible, and the Ecological Crisis*. Kentucky: Westminster John Knox Press.
UNDP. (2007/2008). Human Development Report, 25.
UNFPA. (2011). *State of World Population*, 94.
Wilson, E. (2006). *The Creation: An Appeal to Save Life on Earth*. New York: Norton & Co.

13 Catalysing climate change action in Nigeria

Moderate anthropocentrism and the African perspective of the cosmos

Ada Agada

Introduction

This chapter presents a variant of the anthropocentric model of the environment, dubbed moderate anthropocentrism, as a viable environmental model for Nigeria and Africa. With references to empirical data, the work establishes the negative reality of climate change in Nigeria. It is established that traditional Nigerian (African) conservative practices are rooted in the African holistic view of the universe. Crucially, the work argues that both anthropocentric and non-anthropocentric models are human epistemological windows on the world, which validate the human-centredness of the anthropocentric model.

While it is true that more people today believe we can indeed talk meaningfully about the notion of climate change as a phenomenon that portends danger for living and non-living nature, there is no absolute consensus about this menace. Therefore, it is appropriate to talk about the politics of climate change. The scientific community may well almost uniformly believe that the natural ecological balance has been disturbed in ways that pose serious danger to human and non-human entities, but there is no such consensus among politicians who ultimately decide whether remedial action is necessary or not and the magnitude of intervention (Giddens, 2008). Climate change sceptics believe that there is really no cause for alarm since the observed climactic imbalance and fluctuations are a permanent feature of the physical environment (Giddens, 2008). A category of sceptics, while admitting that there is indeed a problem, overlooks the gravity of this problem and prefers to focus on what the sceptics consider more pressing problems of humanity, for example, wars, poverty, the Acquired Immunodeficiency Syndrome (AIDS) epidemic, etc.

Nevertheless, awareness about climate change continues to heighten as reality mocks the politics of climate change. Activists, national action committees and agencies, international bodies and sundry interest groups have continued to strategise to keep the agenda to save the earth on the front burner, a task made easy by the stark menace of environmental degradation. Unprecedented flooding, desert encroachment, global warming, acid rains, etc., are all fallout of the destructive exploitation of the resources of the planet. The empirical evidence is so glaring that the United Nations International Panel on Climate Change (IPCC) has taken it upon itself to promote awareness of climate change. In furtherance of its goal of helping reverse the harmful effects of climate change, the UN has incubated such treaty bodies as the United Nations Framework Convention on Climate Change (UNFCCC) and the United Nations Convention on Biological Diversity (CBD).

Response to climate change in Africa has been slow even as the debate over the magnitude of this dark phenomenon reverberates around the planet. While the politics of climate change play out on the international stage, Africa continues to feel the full impact of the phenomenon as ocean surges, desertification, air pollution, heat waves, climate change-induced armed conflict and other negative effects wreak havoc on the African people and their environment. Culture, as has been generally observed, varies from place to place. Cultural multiplicity makes it very difficult for one climate change action model developed in one cultural environment to record substantial success when applied to a different cultural environment. The argument for enlisting purely cultural factors and nuances in the fight to preserve the environment follows from the realisation that there are cultural multiplicities in the world and that incorporating cultural ethos into the framework of the fight against climate change will go a long way in making a success of the global endeavour to preserve the earth and hand it in good condition to succeeding generations.

The philosophy of the environment has emerged as a field of philosophy committed to clarifying concepts and issues associated with climate change and supplying appropriate ethical frameworks for the global campaign against climate change. This field is a very young one, even in the West. The philosophy of the environment is virtually undeveloped in African philosophy. Sophisticated philosophical models of the environment have been developed in the West, such as ecocentrism (deep ecology), anthropocentrism and ecofeminism. In this chapter, I intend to contribute to the growth of the philosophy of the environment as a field of African philosophy by highlighting the basic Nigerian, nay African, perspective of the cosmos and showing how this perspective resonates with a moderate anthropocentric theory of the environment within the fundamentally anthropocentric worldview of Africans. I intend to show that the theoretical resources that support practical action against climate change in the Nigerian environment can be extracted from moderate anthropocentrism, the ethical core of which is an affirmation of the anthropocentric communitarianism proclaimed by the founding fathers of African philosophy, thinkers like Senghor, Nyerere, Azikiwe, Awolowo, Nkrumah, Wiredu, Gyekye, Oruka, Gbadegesin, Onyewuenyi, Ramose, Sodipo, etc. I will weave the case for moderate anthropocentrism into the framework of universal panpsychism and the monistic vision of the universe. The notion of panpsychism will lend coherence to the premise of the interconnectedness of all nature, to which any possible argument in defence of the intrinsic worth of non-human living things and the physical environment must refer directly or indirectly for its validity.

I use the term 'Nigeria' and 'Africa' interchangeably with regard to the cultural life-world of the sub-Saharan Africa. The conflation of the Nigerian experience with the African experience underlines the fact that the former typifies the latter.

Thinkers like Hountondji (1996, p. 60) dismiss the idea of a collective African worldview as a myth. Ciaffa (2008, pp. 127–129) believes that this collectivist inclination betrays a revivalist cultural activism, one that, ultimately, loses sight of the necessity of critical thinking in the philosophical project. If Hountondji is to be believed, then the near universal consensus that African ethics is anthropocentric is false. Thinkers like Hountondji arrive at their mistaken stance because they have conflated ethno-philosophy with African philosophy proper. It has been rightly argued that ethno-philosophy must be seen merely as the foundation of African philosophy (Agada, 2015).

I intend to argue that there is no contradiction in asserting straight/fundamental anthropocentrism and contrasting it with moderate anthropocentrism,[1] while recommending the latter as an appropriate ethical-cum-epistemological tool for climate change action in Nigeria. In defending the admissibility of moderate anthropocentrism as a shift in perspective from the rigid ideological commitment to straight anthropocentrism, I will exhibit the inherent but all too often masked anthropocentric presuppositions and underpinnings of such philosophical models of the environment as ecocentrism and ecofeminism. A commitment to straight anthropocentrism that seeks or seems to deny the possibility of mild forms of anthropocentrism is, accordingly, ideological. I use the word 'ideology' in the sense of a refusal to either compromise or see the possibility of a compromise that waters down straight anthropocentrism. I will establish moderate anthropocentrism as a perspective of fundamental anthropocentrism compatible with the holistic worldview of traditional African society. If no theoretical model of the environment escapes the tentacles of anthropocentrism in one form or the other, then a moderate anthropocentrism will be viable as that philosophical model that avoids the radical ideological reaction of anti-anthropocentric models while at the same time broadening the horizon of existence for the fullest possible realisation of the interconnectedness of all things in the universe of matter and energy, an interconnectedness whose ultimate justification is to be found in the notion of panpsychism.

The impact of climate change in Nigeria

The atmosphere has its natural gaseous components like vapour, carbon dioxide, and methane. These naturally occurring gases hardly pose any threat to the environment and living organisms in this same environment as the natural cycle regulates their release and their removal. However, anthropogenic activities, easily measured by the destructive exploitation of the resources of the planet, have sharply contributed to the concentration of green house gases (GHGs), chlorofluorocarbons (CFCs) and hydrochlorofluorocarbons (HCFCs). These emissions have increased the phenomenon called global warming. Scientists estimate that the earth's temperature has risen by 0.4°C in the last 100 years (Idowu et al., 2011, p. 145). Global warming is the direct result of the depletion of the ozone layer (a region of the atmosphere formed by the interaction of sunlight and oxygen), which protects the earth from the sun's harmful ultraviolet radiation. Climate change, then, is the instability that follows the disruptions to the normal atmospheric conditions traceable chiefly to human agency (Idowu et al., 2011, p. 145; Caney, 2015, p. 376). Scientists estimate that by 2030 the heat-trapping artificial compounds in the atmosphere mentioned above could cause temperature to rise by as much as 3% (Idowu et al., 2011, p. 146). The immediate consequences of global warming around the world include irregular rainfall, heat waves, polar ice cap melting leading to ocean surges, deforestation, desertification and insecurity, to mention but a few.

Rapid industrialisation and increased technological sophistication have all but pitched humanity into a toxic black hole of suffocating emissions, with serious, if not calamitous, long-term consequences for living and non-living things. It is estimated that carbon dioxide levels in the environment have risen by about 30% since the Industrial Revolution while methane levels doubled and nitrous oxide levels rose by up to 15% (Ladan, 2014, p. 178). The negative impact of climate change in Nigeria is undeniable. From the Sahel region in the core North to the lush guinea savannah of the Middle Belt and from the forest region of the South to the mangrove swamps of the Niger Delta, the violent disruptions of the natural ecological balance have left visible scars on human life and the physical environment. The Lagos coastal zone comprising Banana Island, Lekki Peninsula, Ikoyi Foreshore, Five Cowries Creek and Victoria Island has been at the mercy of heightened surges of the Atlantic Ocean and unduly heavy rainfall recently (Aledare, Olayiwola and Olaseni, 2014, p. 16).

While heavy rainfall in the Southern parts of Nigeria has compounded the problem of flooding and gully erosion, the core Northern parts of the country have grappled with reduced precipitation and the encroachment of the Sahara Desert. Climate change has been blamed for the rapid shrinking of the Lake Chad which had for a long time been the major hub of agricultural activities in the North-Eastern tip of Nigeria (Ladan, 2014, p. 180). The problem of desertification in the far North is a core factor in Nigeria's current herdsmen-farmers crisis, which has seen nationwide bloody conflicts between farmers and the wandering Fulani herdsmen leading their cattle southward in search of lush grass. The September 2012, floods caused by heightened downpours were particularly devastating in the states along the paths of the River Niger and River Benue, with entire communities sacked, submerged and uninhabitable for weeks (Ladan, 2014, p. 180).

Just as humans have been hard hit by climate change, so have non-human living organisms and the physical environment suffered. Environmental degradation has gone *pari passu* with oil exploration in the Niger Delta region. This region, with its difficult and intricate terrain, is home to a wide and fascinating range of biotas. With an estimated 450-km coastline, the Niger Delta holds the distinction of the world's second largest delta formation (Uyigue and Agho, 2007). Nearly 2,370 square km consists of creeks, rivers, and estuaries; mangrove swamps cover some 1,900 square km of the Niger Delta, which boasts the world's richest wetland (Uyigue and Agho, 2007). The states that make up the Niger Delta are Rivers, Bayelsa, Akwa Ibom, Cross River, Delta, Imo, Edo, Ondo and Abia. The bulk of Nigeria's crude oil reserve is in the Niger Delta. Consequently, the region has experienced the most severe environmental degradation over the decades. Oil exploration on a massive scale has contributed to the phenomenon of acid rains caused by high concentration of carbon dioxide and nitrous oxides in the atmosphere, the destruction of aquatic life caused by oil spills (Amosu et al., 2012, p. 207), erosion of zinc-plated galvanised sheets by accumulated carbonic and nitric acids that compound the housing problem, and the steady loss of biodiversity (Uyigue and Agho, 2007).

Cross River State exemplifies the ecological treasure that the Niger Delta is. 35% of the state's 21,560 square km is tropical rainforest, while about 5% is swamp and mangrove forest (Ogogo, Odigha and Aya, 2013, p. 13). The Cross River rainforest is home to more than 1,545 plant species, with six of such new discoveries in Nigeria and four new discoveries worldwide (Ogogo, Odigha and Aya, 2013, p. 13). It is, therefore, no wonder that the International Union for Conservation of Nature (IUCN) recognises the Cross River rainforest as one of the world's great biodiversity zones (Ogogo, Odigha and Aya, 2013, p. 13). The state itself boasts of about fifteen forest reserves, one wildlife sanctuary, and one expansive national park (Ogogo, Odigha and Aya, 2013, p. 13). All this natural treasure is without doubt worth preserving in its own right, for its own sake, and for the good of humans now and in the future. Yet the anthropogenic manoeuvres that have contributed to ozone layer depletion now pose a palpable threat to the biodiversity of Cross River State. It has been estimated that the state lost about 27% of its forest cover between 1991 and 2008 due to indiscriminate logging, bush-burning, farming activities and climate change (Ogogo, Odigha and Aya, 2013, p. 13).

While the denial of climate change has been shown to be baseless, given the empirical data adduced above, Nigeria is yet to launch a climate change action plan comprehensive enough to save the country from the impending environmental disaster. Thankfully, this impending disaster has already announced itself in the noticeable climactic disruptions occurring in our everyday existence. Denial is no longer possible, but action is demanded as an absolute necessity.

Proof that the notion of environmental preservation is not new to Nigerian (African) traditional culture

With the hue and cry over climate change that today reverberates on the international stage, and the African silence over this matter, one will be forgiven for wondering whether Nigerians, and Africans in general, appreciate the very idea of environmental conservation. There is no cause for wonder, as traditional Nigerian culture never at any time isolated the destiny of the physical world from the destiny of humanity (Dukor, 2014). The celebrated anthropocentric African perspective of the cosmos was never so rigid as to rule out the ontological imperative of a holistic view of reality. Therefore, I do not see any contradiction in the obvious fact that the essential African anthropocentrism created space for the affirmation of the intrinsic worth of non-human organisms and non-living things. The holistic viewpoint already presupposes the essential worth of the various components of the organised whole. The use of animals as totems in traditional Nigerian (African) culture is widely reported and familiar even as nature veneration is integral to the practice of African traditional religion (Dukor, 2014). This matter will be philosophically engaged in a later section of this chapter.

Indigenous conservation policies and cultural practices have been observed across sub-Saharan Africa (Hen, 2006; Infield and Adams, 1999; Taringa, 2006). These policies and practices come in the form of religious observances, taboos, restricted or forbidden zones, etc. Jimoh et al. (2012) report the existence of pro-conservation taboos in four communities around the Oban Hill in Cross River that are enforced by local institutions. These taboos include communal rules forbidding the use of chemicals for fishing purposes (enforced by the Ademwa institution), laws against the killing of pythons (enforced by the Obon institution), laws prohibiting the hunting of endangered leopards (enforced by the Nabor institution) (Jimoh et al., 2012, p. 212).

The Mgbe cult is the supreme authority that ensures compliance with the communal rules. The Mgbe forest hosting these conservation sites is considered sacred ground. Whoever breaks the law prohibiting the use of chemicals for fishing is punished with fines by the Chans of Oban Hill. The Chans even boast of an organised traditional conservation society, the Mfameyin Conservation Society, in the Etang axis of Oban Hill (Jimoh et al., 2012, p. 213). As a totem animal, the leopard is the sacred symbol of the Mgbe society. Naturally, leopard hunting is prohibited. Any hunter who accidentally kills a leopard is expected to suspend further hunting for the day and treat the corpse of the leopard with a level of dignity comparable to what a human corpse deserves. The bile is removed and buried as an appeasement ritual with the public standing as witnesses. The sale of the body parts of the leopard is strictly forbidden, but the meat can be shared among various households (Jimoh et al., 2012, p. 213). Other community-protected animals include the yellow-back duiker, boars, monkeys, elephants and even snails.

Ayotunde and Ada (2013, p. 22) report that about 31.2 hectares of forestland and riverine areas with their land and aquatic resources were successfully conserved by the Ugep people of Cross River using traditional methods, a confirmation of the eloquent submission of Ikuenobe (2014) that environmental degradation is a heritage of colonialism, having never been a problem before the coming of the colonialists with their bifurcating view of reality informed by the mechanistic vision of Newton, Descartes, Kant and other Western thinkers. The conservation sites usually serve religious purposes. They can be worship grounds, location of shrines, masquerade houses and burial grounds of traditional rulers and priests. Community solidarity and consensus is key to the success of the traditional approach to environmental protection. This is the case since fear of the gods and possible divine retribution supplies an important motivation for the observance of the entrenched taboos. In Obubra, the Ohana people prohibit logging in the vicinity of sacred streams. The law against logging helps shield sacred grounds from profane gaze. In the process of protecting local deities (invariably symbolised by inanimate objects like stones or animate things like trees), the people of Adun Okum and Osopong encourage tree-planting activities (Ayotunde and Ada, 2013,

p. 24). More forest cover means more cover for sacred grounds and objects. This, of course, means better protection for the environment.

The environment itself is a complex interactive network consisting of humans, animals, living and non-living things (Chukwunonyelum, Chukwuelobe and Ome, 2013, p. 63). That climate change today has been traced largely to anthropogenic activities shows that the environment has not been accorded due respect. Contempt for the environment or at least a certain indifference to its intrinsic value is inconsistent with the holistic, therefore reverent and welcoming, worldview of Africans now under scrutiny. Among the Igbo of Southern Nigeria, the earth (ala or ani) is regarded as sacred and conceived in panpsychic terms, as indeed is the general African conception of nature (Dukor, 2014). Destructive exploitation of the earth's resources is discouraged. Evil deeds carried out in the Igbo community are regarded as crimes against the earth (Chukwunonyelum, Chukwuelobe and Ome, 2013, p. 64) conceived not merely as a universal mother to which humans return at death but, more importantly, as an element of the sacred, created by God and imbued with His life force (Ikuenobe, 2014). Thus, justice for humans is tied to justice for the physical environment. This reverence for the environment is also pronounced in the Idoma-speaking communities of Central Nigeria which the present author has been privileged to closely observe. Metuh (1991, p. 87) has noted that environmental pollution indicates dirt, the displacement of what is clean and beautiful by what is unclean and ugly. For him, traditional Igbo culture favours the preservation of the physical environment.

The eco-friendly disposition of Nigerian (African) culture: Its roots in communitarian values and the holistic perspective of the cosmos

Mbiti (1969) has rightly located the human phenomenon at the centre of the African universe. This makes African traditional ethics strongly anthropocentric. The human phenomenon is a privileged being that observes nature and uses its resources for its own benefit. This privilege, the ability to observe the physical world, exploit it, and finally rationalise this exploitation of nature is a capacity unique to the human being. The defender of the anthropocentric model of the environment will remind the anti-anthropocentric environmentalist that it is the human being who problematises the question of animal rights and the intrinsic value of physical nature in response to the anti-anthropocentrist's insistence that animals may after all possess rationality by virtue of their capacity to feel pain and their *possible* possession of a language that humans cannot understand. Taking matters further, the anti-anthropocentrist may suggest that our inability to understand the language of animals is no excuse to deny animals rationality. But, again, the anthropocentrist will say that all our perspectives (including the perspective of animal rights and animal rationality) are human epistemic windows. The argument that animals may have their own perspectives is irrelevant since these hypothetical perspectives cannot negotiate the question of animal rights and rationality with our human perspectives. This line of thinking will isolate pity as the ground of the extension of humanity to animals, plant life, and, ultimately, inanimate things. This conclusion is hardly satisfactory, insightful as it is. Universal empathy must itself be philosophically grounded. I will suggest later that the concept of panpsychism supplies such a philosophical grounding.

The Nigerian (African) perspective of the cosmos is holistic as the writings of Nyerere (1968), Senghor (1964), Asouzu (2011), Agada (2015), Chimakonam (2015), and others have emphatically proven. The holistic perspective is an affirmation of the interconnection and interdependence of everything in existence. Asouzu (2011) has used the felicitous term 'missing links' to underline the incompleteness of isolated existents, the very basis of community, and therefore an ontological necessity. For Asouzu reality is a whole: its parts, as missing links, are characterised by an ontological incompleteness that necessitates interdependence. Consequently, it is as necessary for reality to be interconnected and complementary as it is necessary for reality to be or to become.

African monism promotes a robust human-centred ethics which, at the same time, constantly affirms human-environment solidarity. Truly, the human being is at the top of the hierarchy of being on earth, superior to animals, plants, and inanimate objects; yet the human being is answerable to God. One way of reaching God, a *deus absconditis*, is the veneration of animate and inanimate objects in sacred symbolism, as primal forces of nature carrying the imprint of God's immeasurable creativity. African traditional religions would surely not sanction the use of animals as totems and the veneration of inanimate things like streams and forces of nature like thunder if African culture did not regard non-human living organisms and inanimate things as possessing internal consistency and a real worth independent of the human epistemological standpoint. This embrace of nature would then be the basis for regarding lower life-forms and inanimate things as effective objects of veneration and sacred symbolism. I will argue in a subsequent section that implicit in this warm embrace of nature is the assumption of the validity of the proposition: "Everything in existence has a mind and is either alive or potentially alive." This is a panpsychist thesis I will defend in this chapter.

Ubuntu has been celebrated across the sub-Saharan Africa as an anthropocentric, cosmocentric, and theocentric worldview (Chuwa, 2014, p. 13) that underlines community, participation, and equilibrium or harmony. In my opinion, the elucidation of ubuntu, ujamaa, consciencism, negritude in its metaphysical dimension, complementary reflection, consolationism, ezumezu logic, and all other theories of complementation advanced by various African thinkers represents an intellectual struggle to discover metaphysical, logical, and epistemological bases of panpsychism – a grand unified field of African thought. The problem of panpsychism – the question of how seemingly non-living things can possibly possess some form of consciousness or soul and the question of the relation of mind with matter – as a question for philosophy is no longer to be treated with scepticism, for it is one question that must be addressed for us to begin to have a better understanding of mind or consciousness. The denial of mind favoured by eliminativism is merely a way of avoiding the panpsychist question. Panpsychism is increasingly being accepted as worthy of philosophical reflection: so much so that Sprigge (2002,

p. 235) believes that only a supposition of panpsychism can open an epistemic window to the world, a triumphal justification of the age-old African perspective which we now celebrate for its foresight in speaking out for the environment.

An explicit or implicit acceptance of panpsychism is the only basis for affirming the intrinsic value of inanimate things, independent of human subjectivity both in the African and non-African universe. If this assertion is denied, I see no other way forward for animal and environmental rights. The most that can be achieved from a denial is the suggestion that pity is the basis for affirming animal and environmental rights. This is the case because when we plead the inhumanness (cruelty) of animal abuse, for example, we – in fact – merely extend our humanity to animals. Yet, the argument cannot stop with pity, that is, the extension of our humanity to animals. The assumption that we can or should extend our humanity to animals will demand justification, an exercise that will, again, direct us to a panpsychist formulation. Affirming the intrinsicality of the environment while denying panpsychism is possible only through an ideological commitment to liberalism. This commitment merely forbids us from destructively exploiting the environment without demonstrating why we must not assume that non-sentient nature exists for our benefit. Thus, a liberal commitment to environmental protection merely repeats the doctrine of humaneness while pointing beyond itself towards panpsychism without, yet, admitting its conceptual shortcomings.

Before articulating a certain notion of panpsychism within a moderate anthropocentric framework, I will briefly examine the conflict between the anthropocentric approach to the environment and supposedly non-anthropocentric approaches like ecocentrism and ecofeminism. In the process, I will underscore the inescapability of the 'anthropic' viewpoint and the implication for moderate anthropocentrism.

Some ethical models of the environment

The anthropocentric model of the environment is the one most people easily relate with. This is hardly surprising since the primary standpoint of the human phenomenon is anthropic or mind-mediated. The ecocentric model seeks to overlook the weight of the human standpoint and denigrate "human-centred utility" in line with an enthusiastic liberalism that, yet, cannot refute the assertion of the defender of anthropocentrism that all ethical environmental models are mind-mediated. These models are conceptualised based on our experience of the world. This conceptualisation is shaped by our education, beliefs, ideological leanings, opinions, etc. We have no epistemic access to the perspectives of animals and physical nature. Consequently, the granting of intrinsic value to animals and the physical environment is at the pleasure of the human being. Anthropocentrism considers the charge of chauvinism levelled against it and dismisses it precisely on this ground of the primary standpoint. The term "human-centred utility" denotes and connotes the conception of the environment as possessing value only to the extent that it somehow serves human purpose. If we insist that active benefits like logging, mining, oil exploration, etc., are human-centred benefits, we must also be committed to saying that benign benefits like tourism, keeping of pets, and the mere admiration of nature are human-centred benefits to be frowned upon as sternly as the active benefits. The point here is that our attitude to the environment is, from the beginning, inescapably chauvinistic. The very definition of a benefit as something positive accruing to men and women already implicates the epistemic dilemma of every model of the environment. The most passive interaction with the environment is guided by the human standpoint.

How can the environment possess intrinsic worth independent of the human standpoint? The task before the ecocentrist is not merely the spiritualisation of reality, as Grey (1986, p. 214) imagines, but, most importantly, the deployment of a panpsychist argument to justify the intrinsicality of the environment and the ambitious claim of its metaphysical equivalency with the 'chauvinist' human. How this can be possible will be seen in a later section.

The ecofeminist perspective does not fare better in the task of justifying the intrinsicality of the environment. Ecofeminism must go beyond the egalitarian sentiment (noble as it is) and justify the equality claim proposed, ironically, by the human being on behalf of silent nature. Ecofeminism ties the fate of women with the fate of the environment, persuaded that the emancipation of nature has a bearing on the emancipation of women (Gaard, 1993, pp. 1–3). Ecofeminism advocates a holistic approach to the environment, one that dissolves the rigid construction of the self as different (from the other) and embraces an empathetic vision of reality (Birkeland, 1993, p. 20). In pursuing this goal, ecofeminism advocates the destruction of the patriarchal essence which is considered aggressive, bifurcating, and anti-intuitive. As Marti Kheel (1993, p. 257) controversially puts it, "The patriarchal mind has managed to look, but not see, act but not feel, think but not know."

The biocentrism of the ecocentric and ecofeminist perspectives goes down well with the essential African anthropocentric perspective of the cosmos. Both models champion a complementary understanding of reality. However, the environmental fixation of deep ecology that underprivileges the human phenomenon will hardly resonate in Africa. This romantic privileging of the environment over the 'bullying' human being is shared by the ecofeminist model. This utopianism, in my opinion, can be traced to two sources, namely, 1) contemporary Western liberalism and 2) a hesitant attempt to ground the intrinsicality of non-human living organisms and non-sentient nature in a panpsychist conception of the universe. The utopianism of contemporary Western liberalism vis-à-vis the environment question overlooks the masculine or aggressive aspect of non-living nature and the fact that human beings are as much an element of nature as non-humans. Nature here is the totality of things constituting the universe of matter and energy. The perceived aggressiveness of men, which all too easily riles feminists, is very much on display in the sphere of non-sentient nature, thus puncturing the romantic conception of non-living nature as the blameless victim of human manoeuvres. In this feminist context, 'human' means 'man.'

But nature is not blameless. The destructive rage of nature expressed in earthquakes, hurricanes, cyclones, floods, falling asteroids, etc., is too well known. If nature is not blameless, then it is reasonable to uphold an anthropocentric environmental model that acknowledges the concern of other environmental models like ecofeminism and ecocentrism. I propose a moderate anthropocentric model as a shift in perspective within a larger anthropocentric worldview which not only satisfies the human-centred conditions of Nigerian, nay African, culture but also embraces the best aspirations of ecofeminism and ecocentrism, for instance. Moderate anthropocentrism affirms the necessity of holism while escaping the tendency of ecocentrism to over-privilege the environment and ecofeminism to focus on women. Moderate anthropocentrism opens out its arms to all parties in the scheme of nature while privileging the human phenomenon in fidelity to the fact that all standpoints are anthropocentric. I will return to the idea of moderate anthropocentrism in a later section of this chapter.

The attempt to ground the intrinsicality of non-human living organisms and non-sentient nature is without doubt a difficult task. The very attempt throws up difficult questions in metaphysics, epistemology, science, philosophical psychology and the philosophy of religion. Yet for the privileging of the environment not to be consigned to the sphere of mere ideological jingoism (especially in the light of the aggressiveness/destructiveness of nature) a theory that philosophically demonstrates the actuality or possibility of non-sentient nature possessing an internal consistency becomes necessary. I see no other path to the affirmation of the intrinsicality of non-human living organisms and non-sentient nature than the way opened up by panpsychism.

A possible path through panpsychism

Analytical philosophers in the West have written copiously in defence of panpsychism in recent years (Seager, 1995; Strawson, 2009; Chalmers, 1995). The basic premise of panpsychism is that worlds exist because they can be experienced. Experience is essentially subjective, even as consciousness grounds subjectivity. The problematisation of panpsychism in the Western tradition revolves around the emergence of consciousness in a world conceived by science in physicalist terms. Is consciousness reducible to brain states and the neural network and therefore the physical, or is it something irreducible, a fundamental property of matter or even something prior to matter? How does the proposition that mind or mental stuff is present everywhere in the universe fit into the scheme of common sense which clearly abhors the idea that my ballpoint pen has a mind, however dim? As a doctrine, panpsychism is not new to Western philosophy. Forms of the doctrine can be extracted from the thought of the ancient Stoics, Spinoza's infinite causal system, and Hegel's absolute idealism. Even the dualistic system of Kant (2005, pp. 156, 165, 171) repeatedly flirted with the notion of panpsychism.

The growing interest in panpsychism in the West is, in my opinion, due to the fact that this notion of reality is the logical conclusion of the commitment to monism if the world is to make sense in an immanent teleological dimension. I am persuaded that the denial of panpsychism takes the sails out of the theory of evolution, the notion of biogenesis, and belief in the existence of God (Agada, 2015).

Panpsychism is compatible with African culture and worldview. Traditionally, Africans believe that plants and inanimate things have a life-force which justifies what has been contemptuously but ignorantly dismissed as idolatry. Njoku (2014, pp. 13–21) took matters up to the philosophical level when he suggested that the concept of *chi* constitutes the unifying principle in Igbo metaphysics. He suggests that *chi* is a category of being, a principle encompassing destiny and individuality. The idea of a spiritual essence as exemplified by *chi* is shared by most West African ethnic groups (Njoku, 2014, p. 14). For instance, the present author's Idoma ethnic group found in Central Nigeria have their *chi* equivalent in the idea of *owo*. *Owo* as a principle can mean 'god' or 'spirit' or 'guardian angel.' Every human being has his or her *owo*. Can inanimate things have *owo*? The answer is in the positive. God (Owoicho) in Idoma actually means 'god of the sky.' God, then, is supreme as the ultimate principle of nature. It is safe to say that inanimate things have their own gods or guiding spirits or minds. When we reconsider *owo* in the sense of spirit or consciousness, as we move away from the sacred discourse towards the sphere of naturalism, *owo* becomes synonymous with mind and consciousness. Thus, there is mind in all things in varying degrees, although it is in the human phenomenon that mind has developed into conscious subjectivity, such that not only does the human being feel and think but this being also reflects on its feelings and thoughts in what we must understand as individualisation and further humanisation, the making of the unique self.

Thus we see here a valid justification for the African privileging of a human being over a tree or a chimpanzee. Everything in existence has *owo* or mind, but it is in human beings that *owo* has attained the highest, most remarkable, development. Still, we must take the question of panpsychism up to the most sophisticated intellectual level possible in a way independent of Western thought, drawing inspiration from African cosmology and cosmogony to create a philosophy universally applicable. I will attempt a philosophical justification of the intrinsic worth of non-humans without prejudice to moderate anthropocentrism within the framework of consolation philosophy.

The possibility of panpsychism from the perspective of consolation philosophy and the implication for all models of the environment founded on a holistic vision of the universe

Consolation philosophy is an African idealistic philosophy to the extent that it proposes *mood* as the fundamental nature of reality. *Mood* is not merely an affect or a disposition to the world we experience, but, deeper still, the prototype of mind in conscious beings like humans which causes an affect and undergirds all subjectivity actionable up to the most abstract intellectual level. The orientations of *mood* find outlets in desires, feelings and reasons, at the bottom of which subsist the dominant components of *mood* in the human phenomenon: joy and sadness. Reason, then, is subject to *mood* as much as emotion. In a non-living thing, mind has no orientation, although it subsists as this thing's fundamental nature, as a dormant *mood*. Human existence, for instance, has meaning only to the extent that *mood* is active as joy or sadness. If sadness is active at any point, a movement towards the state of joy is initiated, which movement may be emotional or intellectual but which is always motivated by a primal emotional need to access joy whose state of intellectual awareness is intimate with optimism. Failure to actualise this movement consolidates pessimism and the nihilistic view of reality.

In the work *Existence and Consolation: Reinventing Ontology, Gnosis and Values in African Philosophy*, I showed how rationality, the vanguard of reason which is often rightly conflated with reason, is derived from human emotion, which is basic for *mood* since *mood* itself is a composition of joy and sadness at the human level and silence at the level of unconscious beings. I distinguished between the logic of existence or human emotionality and the logic of the intellect, which is our everyday inductive-deductive reason. I argued that the logic of the intellect is reducible to the logic of existence, regardless of the level of intellectual sophistication attained through thought. The proof of this reducibility is the fact that the most abstract kind of thinking is not immune to disputation on the basis of metaphysical preferences and even ideological commitment. A theistic philosopher, for instance, can devise the most ingenious argument in defence of God's existence just as his atheist counterpart can marshal out equally sophisticated arguments against God's existence. The belief on the part of the theist and the non-belief (which ultimately satisfies the conditions of belief in God's non-existence) on the part of the atheist are fundamentally emotions and thus belong to the logic of existence despite the level of abstraction constructed on this basic foundation. I have also shown how the most abstract scientific thinking is reducible to the logic of existence (Agada, 2015).

Thus, consolation philosophy submits that subjectivity is ultimately *mood* and *mood* itself a proto-mind, the fundamental stuff of the universe. This is a panpsychist doctrine. We must go further still to see how the conscious and unconscious worlds can be bridged. The solution is already implied in the foregoing. If *mood* is basic to reality and subsists as an orientation intimate with consciousness, if it is unactuated mind in the unconscious existent, then a bridge has been found that links the human phenomenon and non-human living organisms to non-living but potentially sentient things. The non-living things are potentially sentient because they subsist as *mood*. Therefore, the potential is already actual. If my account is rejected, the experiential world becomes utterly unintelligible, as we cannot even begin to form scientific and non-scientific hypotheses about the origin of life and the physical universe. The theory of evolution and the Genesis account of creation collapse even as we are forbidden to speculate about teleology and mechanism and the issue of the universe being a miracle or a non-miracle (that is, whether the spontaneous emergence of the universe from nothing points to a profound mystery or whether mechanism suffices as an explanation).

The theory of evolution may be silent on the origin of life, but its claims about graduality, mutation and adaptation of species in response to external pressure presuppose abiogenesis. The Genesis account agrees here with evolution, as it announces the emergence of man and woman from dust. Ironically, a radical commitment to abiogenesis leads us to biogenesis, the doctrine that life or something alive begets life. Our universe is structurally complex at micro and macro levels as scientific research into the fine-tuning of the universe has clearly revealed (Collins, 2009). The fact that things grow and evolve, consciously and unconsciously, supports the notion that there is a living impulse implanted in everything. I believe this living impulse to be *mood*: in which case, biogenesis displaces abiogenesis.

However, the outline of consolation philosophy provided here is enough to compel us to entertain the idea of non-living things possessing an internal consistency that qualifies them for respect, independent of the human standpoint, the anthropic necessity. If we are going to avoid mere ideological sentiments and an oversimplifying utopianism in the drive to preserve the environment, only the path opened by an intellectual commitment to panpsychism can lead us to the affirmation of human-independent "fundamental rights" of the environment. Resorting to talks about human chauvinism and egoism serves merely demagogic purposes, since every human perspective of the cosmos, be it pro-environment or anti-environment, is already anthropocentric, already egoistic. The ecofeminist and ecocentric perspectives are no less 'anthropic' than the anthropocentric perspective.

It is easy to identify the chief weakness of the consolationist account of panpsychism. Our solution is an offshoot of the human (epistemological) standpoint, the anthropic necessity, notwithstanding our conviction that this solution logically clarifies the conditions under which a chimpanzee, a mangrove forest, and a waterfall can have 'fundamental' rights. In the metaphysical dimension, doubts remain about the validity of the consolationist solution, as truly as we question any claim about the ontological equality of a woman and a chimpanzee or a tree. If every environmental model gains logical validity from an anthropic or epistemological perspective, it becomes untenable to completely dismiss anthropocentrism, unless we are to be declared rigid ideologues. If moderation is what is required to remove the vexatious matter of human egoism, then moderate anthropocentrism is broad enough to accommodate all radical pro-environment views while remaining faithful to the cause of humanity. Humans are also part of nature, the conscious part of it that has actively realised *mood*. It serves no purpose demonising humans and glorifying non-humans and unconscious objects. That non-humans cannot speak up and defend themselves does not mean they are blameless. When unconscious nature rages and inflicts destruction on humans, for example when an earthquake occurs, we see firsthand the destructiveness of the segment of nature we seek to romanticise.

Moderate anthropocentrism and the Nigerian challenge

The notion of moderate anthropocentrism is intelligible, for it is a shift in perspective within the anthropocentric environmental model to accommodate the concerns over human excesses and the necessity of a complementary interpretation of the universe. Deep ecology's emphasis on the intrinsic worth of all living things (Drengson, Devall and Schroll, 2011, p. 106) finds a higher affirmation in the ecofeminist vision of the interconnectedness of all things (Birkeland, 1993, p. 20). Yet deep ecology's romanticisation of the environment goes against the African privileging of the human. On the other hand, ecofeminism may not gain popular support in the Nigerian setting – which is our primary focus – due to its inclination to a feminising essentialism, the indisputable central feminist focus. Moderate anthropocentrism, like moderate communitarianism,[2] pursues an inclusive agenda while remaining faithful to the human-centred worldview of the Nigerian people who share in the African anthropocentric heritage.

Given the cultural rootedness of the anthropocentric worldview, a radical overthrow of the current order (that is, the dominance of anthropocentrism) to pave way for new, untested, perspectives, will, in my opinion, be counter-productive. In the first place, there is no problem with the anthropocentric perspective. Human-centredness is not the same as 'destructive exploitation.' The eternal legitimacy of human-centred ethics is due largely to the fact that it is the human being that observes the universe and purports to speak for 'silent nature,' whether favourably or unfavourably. The standpoint we know is always a human standpoint, however keen the solidarity and empathy we feel for animals and the physical environment. The human standpoint is already anthropocentric, for it is shaped by our experience of the world. Humans constitute the centre of this experience. Our solidarity with nature is an offshoot of the anthropocentric impulse since this solidarity is an attitude conditioned by wholly human opinions and beliefs. The man who accuses another of being indifferent to the rights of the environment is merely projecting his humanity into the physical world. The nearest thing to a valid justification of this projection of humanity into physical nature (a kind of anthropomorphism) is the foundation that panpsychism offers, a foundation I have so far tried to clarify.

A shift in perspective within the anthropocentric framework that incorporates the communitarian values and holism of the Nigerian people throws up moderate anthropocentrism as a viable model of environmental conservation. In this way, moderate anthropocentrism discards the toga of radical ideological commitment and negotiates an environmental deal with non-anthropocentric models of the environment, whose standpoints are, nevertheless, 'anthropic.'

Conclusion

The reality of climate change can no longer be denied anywhere in the world, and certainly not in Nigeria, where empirical evidence of climate change has become too glaring to be ignored. It may well be that wild fluctuations in climactic conditions and the resultant negative impact on biodiversity and the environment are a permanent historical fixture (Edgerton, 1973), but recent developments call for concern about the way humans handle the environment.

I supplied empirical evidence to underline the severity of the problem of climate change in Nigeria, with particular reference to Cross River State and the Niger Delta. I showed that the practice of environmental conservation is integral to Nigerian and African culture, traceable to the communalism of Africans and their holistic worldview. I argued in favour of a moderate anthropocentric ethical model of the environment as a legitimate and desirable shift of perspective within the overall anthropocentric cosmology of the Nigerian people and pointed to its localisation as an advantage in the struggle to catalyse climate change action in Nigeria. Consequently, the solution to the Nigerian environmental crisis is already available at home. The ethical and ideological components of the fight against climate change are already available. What remains is for the Nigerian government to interface with every community to further strengthen the activities of the Federal Environmental Protection Agency (FEPA). FEPA should consider the necessity of getting the major national and international oil companies to buy into the ethic of moderate anthropocentrism as a viable way of restoring the ecological integrity of the Niger Delta which has been so completely ravaged by the blatant exploitation of crude oil. Embracing this ethic that is at once pro-human and pro-environment will go a long way in improving the Niger Delta environment and promoting peace in this vital region of Nigeria. This way, we can arrive at a healthy compromise between the interest of humans and the interest of the environment considered as possessing intrinsic value.

The epistemological standpoint, our window on the world, is necessarily 'anthropic.' Piety towards the physical environment or non-human living organisms, by virtue of being a psychological perspective seeking epistemological clarification, is *ab initio* anthropocentric. The 'anthropic' is that which follows from the human epistemic standpoint while that which is anthropocentric indicates human-centredness. Our beliefs and opinions are anthropic by virtue of their subjective character. These beliefs and opinions are also anthropocentric by virtue of being shaped by our experience of the world in our capacity as human beings. A pro-environment view that accuses humans of chauvinism is no less anthropocentric than the anti-environment view it disparages, for the former is shaped by events in the world that revolve around the human phenomenon. As I argued earlier, the root of the strongly pro-environment movement in the West is partly in contemporary Western liberalism, which incorporates an ethic of empathy or pity. This ideology-driven commitment is, consequently, a projection of our humanity into non-sentient nature, a hesitant attempt to ground the rights of all beings in panpsychism.

Nevertheless, Africa is not what it used to be before the coming of the colonialists and the dawn of the technological age. While we can look back to our past for inspiration and relearn ancestral piety towards the environment, a return to a primitive way of life in a certain ideological romanticisation of pre-colonial Africa is neither possible nor desirable. Invoking the age-old conservation practices of traditional African communities is an acknowledgment of the complexity of the fight against climate change and the need to mobilise human and material resources at all levels. While we march into the future seeking to crack for ourselves the riddle of science and technology, we can confidently hold fast to the environmental piety of our forefathers, which is rooted in a panpsychist conception of the universe, that all may be well not only with the environment and the current generation but also with coming generations.

Notes

1 I owe Prof Louise du Toit a debt for pointing out this ambiguity.
2 Metz (2014), against Oyowe (2013), has argued for a moderate communitarianism that accommodates human rights within the larger framework of straight communitarianism.

References

Agada, A. (2015). *Existence and Consolation: Reinventing Ontology, Gnosis and Values in African Philosophy*. St Paul, MN: Paragon House.

Aledare, K. D, Olayiwola, K. O., and Olaseni, A. M. (2014). Climate Change Awareness Among Residents of Lagos Island, Nigeria. *American Journal of Social Issues and Humanities*, 4(1), 2–25.

Amosu, A. O, Bashorun, O. W., Babalola, O. O., Olowu, R. A., and Togunde, K. A. (2012). Impact of Climate Change and Anthropogenic Activities on Renewable Coastal Resources and Biodiversity in Nigeria. *Journal of Ecology and the Natural Environment*, 4(8), 201–211.

Asouzu, I. (2011). Ibuanyidanda and the Philosophy of Essence. *Filosofia Theoretica*, 1(1), 79–118.

Ayotunde, E. O., and Ada, F. B. (2013). Conservation of Biodiversity in Central Cross River State, Nigeria: The Role of Indigenous Communities. *Research Journal of Fisheries and Hydrobiology*, 8(2), 20–26.

Birkeland, J. (1993). Ecofeminism: Linking Theory and Practice. In G. Gaard (Ed.), *Ecofeminism: Women, Animals, Nature* (pp. 13–59). Philadelphia, PA: Temple University Press.

Caney, S. (2015). Climate Change. In D. Moellendorf and H. Widdows (Eds.), *The Routledge Handbook of Global Ethics* (pp. 372–386). London: Routledge.

Chalmers, D. (1995). Facing Up to the Problem of Consciousness. *Journal of Consciousness Studies*, 2(3), 200–219.

Chimakonam, J. O. (2015). The Criteria Question in African Philosophy: Escape From the Horns of Jingoism and Afrocentrism. In J. O. Chimakonam (Ed.), *Atụolu Amalu: Some Unanswered Questions in Contemporary African Philosophy* (pp. 101–123). Lanham, MD: University Press of America.

Chukwunonyelum, A. C. K., Chukwuelobe, M., and Ome, E. (2013). Philosophy, Religion and the Environment in Africa: The Challenge of Human Value Education and Sustainability. *Journal of Social Sciences*, 1(6), 62–72. doi: 10.4236/jss.2013.16011.

Chuwa, L. T. (2014). *African Indigenous Ethics in Global Bioethics: Interpreting Ubuntu*. New York: Springer.

Ciaffa, J. A. (2008). Tradition and Modernity in Postcolonial African Philosophy. *Humanitas*, 21(1), 121–145.

Collins, R. (2009). The Teleological Argument: An Exploration of the Fine-Tuning of the Universe. In W. L. Craig and J. P. Moreland (Eds.), *The Blackwell Companion to Natural Theology* (pp. 202–281). Oxford: Blackwell.

Drengson, A., Devall, B., and Schroll, M. A. (2011). The Deep Ecology Movement: Origins, Development, and Future Prospects (Toward a Transpersonal Ecosophy). *International Journal of Transpersonal Studies*, 30, 101–117.

Dukor, M. (2014). Philosophy and African Salvation. *American International Journal of Social Science*, 33, 133–142.

Edgerton, F. N. (1973). Changing Concepts in the Balance of Nature. *Quarterly Review of Biology*, 48, 322–350.

Gaard, G. (1993). Living Interconnections With Animals and Nature. In G. Gaard (Ed.), *Ecofeminism: Women, Animals, Nature* (pp. 1–12). Philadelphia, PA: Temple University Press.

Giddens, A. (2008). *The Politics of Climate Change: National Responses to the Challenge of Global Warming*. www.policy-network.net/uploadedFiles/Publications/Publications/The_politics_of_climate_change_ Anthony_Giddens.pdf

Grey, W. (1986). A Critique of Deep Ecology. *Journal of Applied Philosophy*, 3(2), 211–216.

Hen, L. (2006). Indigenous Knowledge and Biodiversity Conservation and Management in Ghana. *Journal of Human Ecology*, 20(1), 21–30.

Hountondji, P. (1996). *African Philosophy: Myth and Reality* (2nd ed.). Bloomington, IN: Indiana University Press.

Idowu, A. A., Ayoola, S. O., Opele, A. I., and Ikenweiwe, N. B. (2011). Impact of Climate Change in Nigeria. *Iranica Journal of Energy and Environment*, 2(2), 145–152.

Ikuenobe, P. A. (2014). Traditional African Environmental Ethics and Colonial Legacy. *International Journal of Philosophy and Theology*, 2(4), 1–21. doi: 10.15640/ijpt.v2n4a1.

Infield, M., and Adams, W. M. (1999). Institutional Sustainability and Community Conservation: A Case Study From Uganda. *Journal of International Development*, 11, 305–315.

Jimoh, S. O., Ikyaagba, E. T., Alarape, A. A., and Adeyemi, A. A. (2012. The Role of Traditional Laws and Taboos in Wildlife Conservation in the Oban Hill Sector of Cross River National Park (CRNP), Nigeria. *Journal of Human Ecology*, 39(3), 209– 219.

Kant, I. (2005). *Critique of Judgment*. J. H. Bernard (Trans.). Mineola, NY: Dover Publications.

Kheel, M. (1993). From Heroic to Holistic Ethics: The Ecofeminist Challenge. In G. Gaard (Ed.), *Ecofeminism: Women, Animals, Nature* (pp. 243–271). Philadelphia, PA: Temple University Press.

Ladan, S. I. (2014). An Appraisal of Climate Change and Education in Nigeria. *Journal of Geography and Regional Planning*, 7(9), 176–184. doi: 10.5897/JGRP2013.0405.

Mbiti, J. (1969). *African Religions and Philosophy*. London: Heinemann.

Metuh, E. I. (1991). *African Religions in Western Conceptual Schemes: The Problem of Interpretation*. Jos, Nigeria: IMCO Press.

Metz, T. (2014). African Values and Human Rights as Two Sides of the Same Coin: A Reply to Oyowe. *African Human Rights Law Journal*, 14, 306–321.

Njoku, F. O. C. (2014). The Identity of the Particular: An African Basis for Philosophy, Science and Human Development. In N. C. Ogbozo and C. I. Asogwa (Eds.), *Philosophy, Science and Human Development* (pp. 3–34). Enugu, Nigeria: Snaap Press.

Nyerere, J. (1968). *Ujamaa: Essays on Socialism*. Nairobi, Kenya: Oxford University Press.

Ogogo, A. U., Odigha, O., and Aya, F. A. (2013). Geological Restoration and Climate Change Mitigation Strategies in Cross River State, Nigeria. *International Journal of Research in Applied Natural and Social Sciences*, 1, 13–18.

Oyowe, A. (2013). Strange Bedfellows: Rethinking Ubuntu and Human Rights in South Africa. *African Human Rights Law Journal*, 13, 103–124.

Seager, W. (1995). Consciousness, Information, and Panpsychism. *Journal of Consciousness Studies*, 2(3), 272–288.

Senghor, L. S. (1964). *On African Socialism*. M. Cook (Trans.). London: Pall Mall Press.

Sprigge, T. L. (2002). Idealism. In R. M. Gale (Ed.), *The Blackwell Guide to Metaphysics* (pp. 219–241). Oxford: Blackwell.

Strawson, G. (2009). *Selves: An Essay in Revisionary Metaphysics*. New York: Oxford University Press.

Taringa, N. (2006). How Environmental Is African Traditional Religion? *Exchange*, 35(2), 191–214.

Uyigue, E., and Agho, M. (2007). *Coping With Climate Change and Environmental Degradation in the Niger Delta of Southern Nigeria*. www.credcentre.org/Publications/adaptation_nigerdelta.pdf

14 Zimbabwe's environmental crisis

Questioning ubuntu?

Fainos Mangena

Introduction

In this chapter, I argue that Zimbabwe's environmental crisis since 2000 has always been orchestrated by bad environmental management practices, policies and procedures, as well as the failure by the government and people of Zimbabwe to seriously tap into the much treasured indigenous knowledge systems (IKS) in the way they manage the environment. The imposition of Western sanctions on Zimbabwe by Britain and its allies following gross violations of human rights through political violence could not help matters, as the Zimbabwe government allowed disgruntled war veterans and ruling party supporters to invade white-owned farms leading to wanton destruction of vegetation and the killing of animals found on these farms. This, coupled with the collapse of the economy, which saw many companies closing or downsizing in order to dodge this economic malaise, has had a serious impact on the environment, as the majority of the unemployed youths also started destroying the environment in search of gold and other precious minerals – giving rise to what became known as *chikorokoza* (informal mining).

This, obviously, may have had an effect on climate change, resulting in incessant droughts, the drying of dams and the depletion of water resources, which Zimbabwe and the region have been experiencing in recent years. The above problems call for a serious reflection on the place of *hunhu/ubuntu* philosophy in promoting attitudes that can lead to the preservation of the environment in order to avoid this impending environmental catastrophe. As I defend this position, I am aware of the various criticisms that are levelled against *hunhu/ubuntu* as a philosophy and a theory of praxis. I call to mind the criticisms by Bernard Matolino and Wenceslaus Kwindingwi (2013), who argue that, because of some isolated ugly incidences in South Africa in 2012, *ubuntu* must not be seen as a philosophy that can promote peace and harmony as it has outlived its usefulness. I also call to mind arguments by Leonhard Praeg and Siphokazi Magadla (2014), who think that the Marikana massacres must end any talk about *ubuntu* by the African National Congress (ANC), the ruling party in South Africa. I argue that the criticisms levelled against *ubuntu* by Matolino and Kwindingwi (2013) as well as by Praeg and Magadla (2014) are unwarranted since the incidents cited above have nothing to do with *hunhu/ubuntu* but have everything to do with the failure of *hunhu/ubuntu* actors to embrace this philosophy. In the next section, I am going to give an outline of the causes and effects of the environmental crisis in Zimbabwe as part of the background information.

Environmental crisis in Zimbabwe: background information

Often, when the Zimbabwe crisis is told, emphasis is put on the suffering of the people as a result of the political, social and economic challenges that faced Zimbabwe in the last two decades. Rarely is the crisis extended to the suffering of other life forms or non-human beings. As the English people put it, "until lions have their own historians, tales of the hunt will always glorify the hunter." In this chapter, I try to tell the untold story of the suffering of these other life forms as a result of the failure by Zimbabwe to harness *hunhu/ubuntu* in order to live in harmony with nature.

As an entry point to this discussion, I argue that "Zimbabwe's land reform programme which started in earnest in the year 2000, saw desperate land seekers among them war veterans and supporters of the Zimbabwe African National Union Patriotic Front (ZANU PF) invading white owned farms and forcing these white commercial farmers to vacate the farms" (Mangena, 2014, p. 231). Some of these commercial white farmers who tried to resist these invasions were beaten up, maimed and killed and the rest fled to neighbouring countries such as Mozambique, Zambia and South Africa to seek refuge (Mangena, 2014, p. 231). This move had a serious impact on the management of the environment as these new farmers started to destroy forests as well as killing animals which were being reared by the commercial white farmers. Despite calls by animal rights groups such as the Society for the Prevention of Cruelty to Animals (SPCA), these land invaders kept on killing animals with reckless abandon.

They took whatever these white commercial farmers had produced and, "they started slaughtering them *en masse* . . . (2014, p. 232). As De Georges and Relly (2007) remark:

> Besides the abuses of animals in the invaded farms, the government . . . appeared to have permitted poaching in national parks, as the rural populace started killing elephants and other game animals for meat in an operation code named 'Operation *Nyama*'.
>
> (Meat)

This, together with the human rights abuses involving the maiming and killing of opposition supporters, resulted in America and its allies imposing sanctions on Zimbabwe which, up to this day, have had a negative impact on the performance of the economy leading to high unemployment levels, poverty and the general lowering of the standards of living. Unemployment forced many youths to engage in illegal mining activities and these had an effect on environmental management, especially as this search for gold and other minerals led to environmental degradation. Chimonyo, Mungure and Scott (2013, p. 23) observe that "the discovery of diamonds in Marange (Chiadzwa) worsened an already dire situation, as diamond mining in Chiadzwa started as a free for all activity where miners and traders, middlemen and the security institutions co-existed."

The effect of these activities is being felt in Zimbabwe today, with incessant droughts being the order of the day as well as the drying of dams such as Mazowe dam in Mashonaland Central, which used to supply water in the province's Mazowe Citrus Fruits Estate. The water levels in the Kariba dam, which is located in Mashonaland West, and which generates hydro-electric power for Zimbabwe, has also been a cause for concern, coming at a time when most cities and towns are always experiencing massive load shedding. Water pollution has also resulted in unsafe water supplies from such dams as Lake Chivero, thereby causing water-borne diseases such as cholera and typhoid, as well as non-communicable diseases such as cancer because of some of the cheap chemicals which are used to purify this water. All these problems are a result of the government's failure to harness *hunhu/ubuntu* and promote an environmental harmony in which people treat the environment and its contents as important as they are. Below, I discuss the role of *hunhu/ubuntu* in the promotion of environmental harmony in Zimbabwe.

Hunhu/Ubuntu and the environmental crisis in Zimbabwe

In this section, I define and characterise the discourse of *ubuntu* by showing its nexus with environmental philosophy. To begin with, the term *hunhu/ubuntu* is derived from the Bantu languages of Southern Africa, referring to the idea of community or shared humanity (Boon cited in Prozesky, 2009, p. 301). The Nguni expression *Umuntu ngumuntu ngabantu* (Zulu/Ndebele/Xhosa) – which has its Shona equivalence as *Munhu muvanhu* (a person is a person through other persons) – best explains the idea of community or shared humanity in Southern Africa. Prozesky (2009, p. 301) remarks that:

> With this principle deeply embedded in the African consciousness, it follows that Western style individualism is judged as a dangerous error bringing great harm to human communities, and elevating self-gratification above concern for the common good.

For Prozesky, the idea of the common good is much more important than the emphasis on individual rights and freedoms, and, as we shall see later, this idea is best explained by *ukama* (relatedness) as defined and characterised by Munyaradzi Felix Murove (2009). It is also important to observe that the idea of elevating community ahead of individuality has been challenged by some *ubuntu* scholars such as Michael Onyebuchi Eze (2008). For Eze (2008), while many *ubuntu* writers have tended to glorify community over and above the individual, it is important to note that "the individual and the community are not radically opposed in the sense of priority but engaged in contemporaneous formation." In my critique of Eze's argument, I argue that, it is not possible for these two concepts to be "engaged in contemporaneous formation," because the relationship is such that the individual is submerged within the community and the interests and aspirations of the community always matter more than those of the individual (Mangena, 2016b). As put by Placide Tempels:

> The Bantu cannot conceive of . . . the human person as an independent being standing on his own. Every human person, every individual is, as it were, one link in a chain of vital forces: a living link both exercising and receiving influence, a link that establishes the bond with previous generations and with the forces that support his own existence. The individual is necessarily an individual adhering to the chain.
>
> (Tempels cited in Bujo, 2001, p. 113)

The above submissions by Tempels clearly show that community is prior to the individual and the individual's existence makes sense only if his/her contribution is there to further the interests of the community of his or her present generation as well as the interests of the generations gone by and those yet to come. Whatever this means, the point is that the individual is more dependent on the community than vice versa. For example, when the individual has acquired some wealth or some education, he or she has to use the wealth or education to improve the welfare of his or her kin, especially immediate family members. This is important because the wealth or education was acquired to benefit the group in the first place.

A person who does not share his or her wealth or material resources is regarded as *munhu asina hunhu* (a person who is morally deficient) and such a person risks being isolated by the community should a misfortune befall him or her. If he or she loses a child or spouse through death, community will say: *Pfuma yake yake ndiyo ichamubatsira kuchema mwana wake kana mudzimai/murume wake kana afa* (his or her wealth will help him or her mourn the loss of his or her child or spouse). The point is that selfishness has serious repercussions on individual standing and well-being.

For Mangena (2012, p. 10), the concept of *hunhu/ubuntu* is dialogical, involving *vanhu/abantu* (people), *midzimu/amadlozi* (ancestors) and *mwari/unkulunkulu* (God). The elders are responsible for the formulation of *hunhu/ubuntu* moral principles, which are enforced by the spirit world, and the process of arriving at these principles requires that the elders collectively agree on what is good or bad, based on their experiences and collective wisdom. This collective agreement, which is packaged in proverbs, idioms, folktales and riddles, is one which I call the Common Moral Position (CMP). The CMP is passed from one generation to the other and it constitutes the moral code of the people of Southern Africa, in general, and the Shona people of Zimbabwe, in particular.

To understand the role and place of *hunhu/ubuntu* in the promotion of environmental harmony, it is important to look at the concept of *ukama* (relatedness) as expounded by Murove (2009). As Murove (2009, p. 316) correctly puts it, "*ukama* is a Shona word implying relationship and an understanding of reality in terms of interdependence." This kind of relatedness and interdependence, for Murove, goes beyond marital and blood ties as well as species boundaries (2009, pp. 316, 323). He remarks thus:

> An ethic of *ubuntu* is an ethic of holism. It is holistic in the sense that my well-being as an individual has its ontological meaning in relationship to my fellow human beings, those that are still alive, ancestors, God and the environment. From this point of view, one can argue that while *ukama* is a communal ethic, the ecological dimension enshrined in it surpasses anthropological communalism.
>
> (Murove, cited in Prozesky, 2009, p. 302)

Murove maintains that "through *ukama*, the community's moral values are conveyed to individuals from childhood into adulthood" (2009, 316). What this means is that children are taught the importance of living in harmony with each other as well as living in harmony in nature. Through this kind of socialisation, the CMP is hammered into the minds of these children. As Vimbai Gukwe Chivaura (2013) would put it: *Tinofanira kuita hunhu hwakanaka nekubatana nekuti tiri hama ne hama nezvakatikomberedza* (we ought to exhibit good characters and be united because we are related to nature).

Murove maintains that the idea of *hunhu/ubuntu* as a philosophy and an ethic loses its lustre once it is no longer married to the idea of *ukama* because "to be fully human is to be in *ukama* (to be related) with all that constitutes existence" (2009, p. 323). Murove argues that, unlike in the West where the idea of *ukama* between human beings and nature is no longer existing because of technological advancement, the African notion of *ukama* is something to be cherished and valued (2009, p. 327). It is also important to remark that the CMP is the glue that makes it possible for *hama* (relatives) to live in harmony with each other and with physical nature. To this end, Benezet Bujo (2009, p. 289) argues thus:

> The first world should become literate once again and learn how to interpret nature's symbols. Dialogue with the non-Western world should take place and countries should be warned not to demonstrate dominion over all creation in such a deadly manner.

What both Murove (2009) and Bujo (2009) seem to be suggesting here is that the West must abandon what I would call "the destructive tendencies of anthropocentrism"[1] and learn that human beings and physical nature need each other. Thus, Murove (2009) and Bujo (2009) are pointing to the interconnectedness and interdependence that exist in Africa between human beings and nature. To them, the fact that nature participates in human affairs, and vice versa, is a sure sign that the idea of interconnectedness and interdependence is the essence of *hunhu/ ubuntu* philosophy and ethics, as opposed to Hobbesian philosophy, deontology and teleology as celebrated in the West, which valorises individualism. The question to ask is: How does nature participate in human affairs?

There is so much philosophical literature produced by African scholars which shows that nature participates in human affairs. In this chapter, I rely on the literature produced by Bujo (2009), Nisbert Taringa (2014) and Zvakanyorwa Wilbert Sadomba (2014). For Bujo, "cosmic elements enable community life to function. The appearance of the new moon, for example, is always a happy occasion for a village meeting, which through dancing and other activities consolidates community relationships" (2009, p. 291). Taringa cites many examples where nature has always participated in human affairs in Shona society. In this chapter, I cite just one of his numerous examples as follows:

> When a traveller saw a baboon or baboons seated or walking, that was a sign that the journey was safe and it also meant that the ancestors had blessed the journey. However, a baboon seated with its palm on the cheek looking sad signified trouble, that is, it meant that the journey was unsafe and that the ancestors had not sanctioned it. In the event that the journey was urgent and could not be cancelled, the traveller was advised to perform the ritual of putting snuff under a tree clapping hands and pleading for guidance during the journey. Thus, baboons played a dual role, that is, they were bearers of both good and bad news.
>
> (Taringa, 2014, pp. 251–252)

From this, it is clear that nature, or the environment, cares about the safety and welfare of human beings. For Sadomba, "indigenous Zimbabweans have a cosmology that is different from that of Europeans. This cosmology is based on a philosophy that recognises the harmonious trinity of nature, society and the spirit world" (2014, p. 352). This relationship, for Sadomba, is made possible by the communication that exists between the material world and the spiritual world through media that include people, flora and fauna (2014, p. 353). So, whichever way one may want to look at it, nature participates in human affairs and this participation brings to bear the idea of interdependence.

As we have learnt from Murove that the concept of *ukama* should put emphasis on the idea of interdependence, it is also expected that human beings must also care for the safety and welfare of nature or the environment and its contents. To this end, Mogobe B. Ramose (2009, p. 309) argues that:

> To care for one another, therefore, implies caring for physical nature as well. Without such care, the interdependence between human beings and physical nature would be undermined. Moreover, human beings are indeed, an intrinsic part of physical nature although possibly a privileged part.

Looking at the environmental crisis facing Zimbabwe today because of the activities of human beings, it is difficult to argue that human beings care for the safety and welfare of physical nature. It is also difficult to argue that human beings value *ukama pakati pavo ne nharaunda* (the relationship between them and nature). The question then is: Why are the Shona/Ndebele people behaving that way and yet they are inclined to act in a manner that promotes environmental harmony because of the idea of *ukama* which is inherent in them?

The first possible response to the above question could be that because of modernity, which emphasises individual rights at the expense of *ukama*, the Shona/Ndebele people have been caught in a web of ambivalence where there is need to cherish the values of the past that have kept them in harmony with nature, from time immemorial, and at the same time wanting to exert their independence (instead of interdependence) as individuals with fully fledged rights and privileges. This kind of ambivalence has resulted in the creation of multiple identities, especially in Africa, where one wants to embrace both Western and African values at the same time, with the latter always at the tail end of priorities. But whichever way one may want to look at it, the actions of these indigenous people will always be judged based on the values of *hunhu/ubuntu*, especially the attendant concept of *ukama*, which emphasises the importance of relating with physical nature, as put by both Murove and Chivaura.

The second possible response to the above question is that, since 2000, the Shona/Ndebele people have probably abandoned their *hunhu/ubuntu* values because they have seen these values as threatening their survival, amid a sea of poverty caused by the collapse of the country's economy. Thus, an argument can be made that the problems leading to Zimbabwe's environmental crisis have nothing to do with the generality of the population, who have sought survival strategies through whatever means that are available. The problem is with the government of ZANU PF, which, through its bad policies, has thrown people into abject poverty. Although this argument sounds persuasive at first glance, it is difficult to sustain because it gives the impression that when one's survival instincts are threatened or are under attack, he or she has a right to employ unethical strategies to survive.

I argue that no amount of suffering can justify unethical behaviour. Every human being *(munhu/umuntu)* must be guided by the dictates of the CMP as underpinning the philosophy of *hunhu/ubuntu* (as enshrined in *ukama*) in whatever they do. Human beings should always know that they need other beings to exist and, as such, they should desist from wantonly destroying forests and recklessly killing wildlife. The third possible response to the above question is that Zimbabweans probably felt let down by their government following the land invasions of 2000 to 2001, especially given that these invasions had led to the disruption of farming activities, which, in turn, worsened an already dying economy.

They probably felt that the political violence targeting mostly opposition supporters, poverty, the increasing rate of unemployment, the fall of the Zimbabwe dollar, hyperinflation and the plummeting health delivery system and education standards had more to do with bad governance. It is possible that they had pent-up emotions which led them to disregard any talk of values, especially by a government which they had considered to be the source of their problems. In their thinking, the government of Zimbabwe lacked the moral legitimacy that is required to gain the respect of the governed and so they (the generality of the population) were free to do whatever they thought would give them food on the table, and this ranged from poaching, gold panning, diamond mining as well as water pollution in search of those precious minerals.

Assuming that Zimbabweans had made a good case against their government, considering the strength of the above premises, can their actions escape moral criticism? My position is that their actions cannot escape moral criticism in the sense that there were better ways of registering their disappointment with the government than destroying the environment, which will in future, not only benefit them, but even generations to come after them. Besides, the actions were wrong because they disregarded the importance of *ukama* as it relates to human beings and nature. In order to address the problem of the environmental crisis facing Zimbabwe today, there is a need to invoke the spirit of *hunhu/ubuntu* in the way Zimbabweans interact with physical nature, that is, there is a need to take stock of the stipulations of the CMP which undergirds the philosophy of *hunhu/ubuntu*. The question that begs for an answer, though, is: Is *hunhu/ubuntu* as a philosophy or ethic immune to criticism to be appealed to as a solution to Zimbabwe's impending environmental crisis? In the next section, I try to respond to this question.

The limits of *hunhu/ubuntu* philosophy in averting Zimbabwe's environmental crisis

There have been consistent voices in the discipline of philosophy, particularly in Southern Africa, which have dismissed the power and force of *hunhu/ubuntu* as a social, political and economic philosophy worth celebrating. These philosophers range from those born and bred in the continent and those based in Africa but not native Africans (European Africans) to those outside of Africa (armchair critics). In the first category of philosophers, I have in mind the likes of Bernard Matolino and Wenceslaus Kwindingwi (2013), who have argued that the lack of social cohesion in South Africa is a sure sign that *ubuntu* as a narrative of return and a philosophy has become irrelevant and out of touch with reality. They cite the following example to defend their thesis:

> At the beginning of 2012 South Africa was stunned by a television footage that showed an extreme act of barbarity meted out by a taxi driver to an apparently innocent pedestrian. Bloemfontein CBD CCTv footage that was screened on major television news bulletins captured the driver of the said taxi making a hasty reverse at a set of traffic lights. He diverted his vehicle from the road and darted onto the pavement where he knocked a pedestrian down. He reversed and alighted to check the impact of his unique driving skills. Upon realising that his vehicle had not exerted considerable damage to the pedestrian, the driver unleashed a powerful kick to the abdomen of the poor pedestrian who was struggling to raise himself to his feet . . . Other healthy-looking males standing around this shocking scene appeared either unbothered or too scared to come to the aid of the hapless pedestrian.

> The Minister of Police appeared on the News Channel Africa television to bemoan the lack of *ubuntu* in reaction to this incident.
>
> (Matolino and Kwindingwi, 2013, p. 197)

There is no doubt that the behaviour of the taxi driver bordered on nothing but just malice, barbarism and primitivity especially as this happened in an independent South Africa. Matolino and Kwindingwi are justified in describing this behaviour as "shocking." There are many other cases of barbarism that South Africa has recorded in recent years, for instance, the xenophobic attacks of 2008 and 2015, which have put South Africa in the spotlight for the wrong reasons. But to argue that, because of these ugly incidents, *ubuntu* has failed is to miss the point completely. The power and force of the law is not lessened by its continual violation or breach. Law remains law even if people continue to violate it. Does the fact that people continue to commit rape in South Africa and elsewhere in the world lessen the effect of the legal instrument that outlaws rape? Inasmuch as the problem is not with the law but the violators of it, I think the same applies to *ubuntu*.

At this juncture, I will look at how Matolino and Kwindingwi (2013) discredit *ubuntu* as a narrative of return. To begin with, Matolino and Kwindingwi (2013) argue that the continent has been greatly compromised by an appeal to narratives of return which, according to them, have yielded nothing. They cite the examples of negritude (Senghor 1964), ujamaa (Nyerere 1968), consciencism (Nkrumah 1964), Mobutuism (Callaghy 1980) and humanism (Kaunda 1966) as narratives that simply died without bringing a solution to the human condition. Matolino and Kwindingwi (2013) also argue:

> On the political front the search for the narrative of return led to the suppression of political space, tyrannical dictatorships developed in pursuit of a one party state that was supposed to be an embodiment of the people's aspirations.

It is important to note that the problems cited by Matolino and Kwindingwi above do not point to the irrelevance of *ubuntu* in Southern Africa in general and in South Africa in particular; it has more to do with *ubuntu* actors,[2] that is, individuals who have to live or be guided by the stipulations of *ubuntu*. As I noted earlier, a moral requirement does not lose its relevance because some people have decided to violate it. A moral requirement will remain a moral requirement even if people decide not to be guided by it. To this end, Metz (2014, p. 66) uses evolutionary theory to show that the positive effect of moral requirements is not lessened by the fact that human agents are violating them or have chosen not to be guided by them. Metz argues that there is a category of evolutionary theorists called Social Darwinists who believe that the state has no mandate to help the poor. In his response to Social Darwinists, Metz argues that we cannot completely disregard the whole evolutionary theory because some defenders of it have weird ideas. Metz compares this with the idea of *ubuntu* when he remarks that, "even if it were shown that some who believe in *ubuntu* or who appeal to the word *ubuntu*, act wrongly . . . it would not follow that *ubuntu* theory . . . is false or epistemically unjustified."

Two points can be extracted from Metz's argument. Firstly, *ubuntu*, as a theory which is relevant to the needs and aspirations of contemporary South Africa, remains invaluable. Secondly, it is important to observe that the pitfalls of human agency cannot be blamed on *ubuntu*. Based on these two premises, it can be argued that Matolino's and Kwindingwi's arguments are faulty. Thus, the situation in South Africa where *ubuntu* is blamed because people have failed to live according to its dictates is synonymous to the situation in Zimbabwe, where people destroy the environment and blame something else for their lack of ethical responsibility.

Ubuntu can still help Zimbabweans to live in harmony with nature, thereby promoting environmental well-being for as long as people can accept being guided by it. It is grossly unfair to trash a philosophy because a few misguided elements have failed to live by its dictates. I wonder how this is different from the act of burning a whole house because a snake has entered into the house. In the second category of philosophers who have trashed *ubuntu*, I have Leonhard Praeg (2014) who also follows Matolino and Kwindingwi (2013) in dismissing *ubuntu* when he argues that the ANC has no right to claim that the new South Africa was founded on *ubuntu*, if the Marikana massacres are anything to go by. Praeg (2014, p. x) avers that the ANC's

> claim to have founded the new South Africa on the *nomos*, the spirit of law, of an extraordinary humanism has bottomed out. After Marikana, no such claim can possibly be made or sustained . . . Any claim by the ANC government to have founded our democratic political-juridical order on the idea of a shared humanity will and indeed must in future be met with derision. For this reason, alone, it would be the most obvious thing in the world to denote, and dismiss, or perhaps embrace this as the dawn of our post humanist present, to abandon talk of Ubuntu altogether, to dismiss it as a historical artifact.

There are three important points to explore in the above paragraph. Firstly, *ubuntu* is blamed for the failures of the ANC. Secondly, there is a deliberate attempt to try and throw away the baby with the bathwater when philosophers deliberately decide to dismiss *ubuntu* on the basis of the failures of *ubuntu* actors to live by its dictates. Thirdly, and just as with the case of Matolino and Kwinginwi, *ubuntu* is trashed or dismissed because of one ugly incident that has shocked South Africa. The impression that is given is that the ANC government has failed to promote the spirit of shared humanity because of the failures of *ubuntu*; and yet the opposite is true that the ANC government has failed to live by the dictates of *Ubuntu*, leading to the ugly incidents cited by Matolino, Kwindingwi and Praeg.

My problem with this kind of reasoning is that, in blaming *ubuntu*, the argument exonerates the actor from any wrongdoing and yet the actor must take full responsibility for his or her actions. It is the same in Zimbabwe, where people who are recklessly destroying the environment always want to blame the government and the "failure" of *ubuntu* to help put food on the table. It is common in Zimbabwe these days to hear people saying: *Hunhu hunodyiwa here? Hunhu hwenyu hwamunongotaura ndohwatiuraya nenzara* (Do we eat *hunhu*? Your constant talk about *hunhu* has caused us this suffering).

With regard to the environmental crisis in Zimbabwe, one will hear the same people who are destroying the environment blaming the government for their suffering. The point is that it does not matter whether people's lives are threatened by hunger or poverty; the point is that *hunhu/ubuntu* values cannot be mortgaged for anything, and it is unfair to blame *hunhu/ubuntu* for the problems caused by those who have failed to live according to its dictates. As Bujo (2009, p. 292) puts it, "humankind is looking for an ethic to ensure that the earth may be permanently habitable." In the case of Africa, this ethic has to be *hunhu/ubuntu. Hunhu/ubuntu* is the only solution to Zimbabwe's environmental crisis, despite many attempts by some of the African philosophers cited above to discredit it.

Conclusion

This chapter challenged Zimbabweans to embrace the spirit of *hunhu/ubuntu* in their daily encounters with nature. The chapter began by defining and characterising the environmental crisis facing Zimbabwe as a result of human beings' reckless behaviour towards the environment. As this chapter pointed out, this reckless behaviour was made more revealing during the land invasions which took place at the beginning of the new millennium spearheaded by war veterans and some supporters of the ruling ZANU PF who, after evicting the white commercial farmers occupying these farms, started killing animals and chopping down trees with reckless abandon. The chapter also argued that the fall of the Zimbabwe dollar and the subsequent drop in the standards of living as a result of the imposition of Western sanctions saw many unemployed youths involving themselves in gold panning and diamond mining in order to eke out a living, with all these activities contributing to land damage. The chapter argued that, as a result of these activities, Zimbabwe is now facing challenges relating to incessant droughts, reduction of biodiversity, unavailability of clean water because of water population, drying of dams and atmospheric pollution. The solution to these problems, according to this chapter, is an appeal to the philosophy of *hunhu/ubuntu*, as this will ensure that there is environmental harmony between human beings and nature. The chapter tapped into the work of Murove which relates *hunhu/ubuntu* to *ukama;* it emphasised the point that there was *ukama* between *vanhu/abantu* and *nharaunda* (human beings and the environment) and so there was no need for human beings to behave the way they were behaving. The chapter also made reference to the works of Ramose, Prozesky and Bujo, who also argued in the same spirit of promoting environmental harmony by respecting nature. The chapter did not turn a blind eye to the criticisms usually levelled against *ubuntu*, as it cited the works of Matolino, Kwindingwi and Praeg, who argue that *ubuntu*, especially in South Africa, has become irrelevant. They cite examples that point to violations of human rights to defend their theses. However, in reaction to these criticisms, the chapter argues that the problem is not with *hunhu/ubuntu*. The problem is with *ubuntu* actors, that is, human beings who have failed to embrace the spirit of *ubuntu* in their lives. *Hunhu/ubuntu*, therefore, remains the philosophy to appeal to if we are to decisively deal with the environmental crisis facing Zimbabwe today.

Notes

1 By "the destructive tendencies of anthropocentrism," I mean homocentric attitudes that undermine the value of other species in the environment which have a negative impact on the well-being of the environment and humans as well.

2 By *Ubuntu* actors, I mean the practitioners of *ubuntu* who are the human beings themselves. These exclude the mentally deranged and minor children, both of whom have no capacity to make moral decisions.

References

Bujo, B. (2001). *Foundations of an African Ethic: Beyond the Universal Claims of Western Morality*. Nairobi: Paulines Publications Africa.

Bujo, B. (2009). Ecology and Ethical Responsibility From an African Perspective. In M. F. Murove (Ed.), *African Ethics: An Anthology of Comparative and Applied Ethics*. Scottsville: University of KwaZulu Natal Press.

Chimonyo, G. R., Mungure, S., and Scott, P. S. (2013). *The Social, Economic and Environmental Implications of Diamond Mining in Chiadzwa*. Centre for Research and Development. Mutare: Southern Africa Resource Watch.

Chivaura, V. G. (2013). *Hukama ne Nharaunda*. Interview held with Dr Vimbai Gukwe Chivaura at the University of Zimbabwe in June.

De Georges, A., and Relly, B. (2007). Politicization of Land Reform in Zimbabwe: Impacts on Wildlife, Food Production and the Economy. *International Journal of Environmental Studies*, 64(5), 571–586.

Eze, M. O. (2008). What Is African Communitarianism? Against Consensus as a Regulative Ideal. *South African Journal of Philosophy*, 27(4), 106–119.

Mangena, F. (2012). Towards a *Hunhu/Ubuntu* Dialogical Moral Theory. *Phronimon: Journal of the South African Society for Greek Philosophy and the Humanities*, 13(2), 1–17.

Mangena, F. (2014). Environmental Policy, Management and Ethics in Zimbabwe, 2000–2008. *Journal of Pan African Studies*, 6(10), 224–240.

Mangena, F. (2016a). African Ethics Through Ubuntu: A Postmodern Exposition. *Africology: Journal of Pan African Studies*, 9(2), 60–74.

Mangena, F. (2016b). Hunhu/Ubuntu in the Traditional Thought of Southern Africa. In *Internet Encyclopaedia of Philosophy*. www.iep.utm.edu/hunhu/ (accessed June 24, 2016).

Matolino, B., and Kwindingwi, W. (2013). The End of Ubuntu. *South African Journal of Philosophy*, 32(2), 197–205.

Metz, T. (2014). Just the Beginning of Ubuntu: Reply to Matolino and Kwindingwi. *South African Journal of Philosophy*, 33(1), 65–72.

Murove, M. F. (2009). An African Environmental Ethics Based on the Concepts of *Ukama* and *Ubuntu*. In M. F. Murove (Ed.), *African Ethics: An Anthology of Comparative and Applied Ethics*. Scottsville: University of KwaZulu Natal Press.

Praeg, L. (2014). Preface. In L. Praeg and S. Magadla (Eds.), *Ubuntu: Curating the Archive*. Scottsville: University of KwaZulu Natal.

Praeg L. and S. Magadla (Eds.), (2014). *Ubuntu: Curating the Archive*. Scottsville: University of KwaZulu Natal.

Prozesky, M. (2009). Well-Fed Animals and Starving Babies: Environmental and Developmental Challenges From Process and African Perspectives. In M. F. Murove (Ed.), *African Ethics: An Anthology of Comparative and Applied Ethics*. Scottsville: University of KwaZulu Natal Press.

Ramose, M. B. (2009). Ecology Through Ubuntu. In M. F. Murove (Ed.), *African Ethics: An Anthology of Comparative and Applied Ethics*. Scottsville: University of KwaZulu Natal Press.

Sadomba, Z. W. (2014). The Application of African Philosophy in Zimbabwe's Land Revolution. In C. G. Mararike (Ed.), *Land: An Empowerment Asset for Africa: The Human Factor Perspective* (pp. 250–265). Harare: University of Zimbabwe Publications.

Taringa, N. T. (2014). Animals as First Class Citizens. In C. G. Mararike (Ed.), *Land: An Empowerment Asset for Africa: The Human Factor Perspective* (pp. 250–265). Harare: University of Zimbabwe Publications.

15 Exploring the African traditional belief systems (totems and taboos) in natural resources conservation and management in Ghana

Francis Diawuo and Abdul Karim Issifu

Introduction

The aim of this chapter is to discover how taboos and totems as belief systems have been used successfully to promote natural resources conservation and management in the Sankana and Tongo-Tengzuk communities of Ghana. Using primary and secondary data sources, it was observed that the communities have served as classic examples of successful traditional natural resources conservation and management through totems and taboos. Therefore, the authors, on the basis of the above success, recommend a holistic approach involving the government through the Ghana Tourism Authority, civil society organisations and religious leaders to encourage people to apply by-laws and other traditional belief mediums that link human existence to the natural environment in the conservation efforts in Ghana.

The current increasing rate of natural resource loss is a major threat to both human and animal survival. The loss of each species comes with the loss of potential economic benefits, as well as loss of ecosystem balance (Attuquayefio and Fobil, 2005). As such, there has been increased interest in issues relating to the environment all over the world. In particular, the international community has taken leading steps in ensuring proper conservation of natural resources through formal and professional standards.

Meanwhile, before the introduction of modern forms of natural resource conservation and management, indigenous African communities often developed elaborate resource management systems (Fabricius, 2004), as had other local communities throughout the world (Ostrom, 1990). Local groups of people managed the land on which they lived and the natural resources they were surrounded by for millennia (Roe, Nelson and Sandbrook, 2009). There existed locally well-informed traditional beliefs that helped in conserving the available natural resources. Daniel Attuquayefio and Sandra Gyampoh (2010) argue that, before the advent of modern natural resource conservation, traditional societies operated a complex religious and cultural belief systems via norms, myths, taboos, totems and closed seasons to preserve, conserve and manage certain natural resources. The use of these belief systems was geared towards protecting and promoting communal wellbeing, rather than individual interests.

Despite the potency and the role of traditional African belief systems in natural resource management and conservation, little attention is given to this informal institution. Although efforts to integrate rural people into the conservation of natural resource programs and projects have been in place for quite some time (Hulme and Murphree, 1999), the integration has been slow. This could be as a result of increasing non-adherence to long-held traditional beliefs, due to the advent of Western technology, the growing influence of foreign religion and beliefs, lack of modern regulations to enforce the traditional rules, and problems of migration, urbanisation and resettlement (Ntiamoa-Baidu, 1995). Following the above assertion, this chapter seeks to explore further how traditional belief systems such as taboos and totems have contributed to the natural resources conservation and management in Ghana. The chapter also seeks to offer pragmatic recommendations on the need to integrate modern laws, traditional customs and norms in the natural resources conservation and management for the benefit of the generation yet unborn.

The chapter is grouped into six sections. The first section is on the research methodology and approach used in collecting data. While section two defines key concepts, section three delves into the debate between Western and African belief systems on Natural Resources Conservation and Management. Sections five and six constitute results, discussions and conclusion, respectively.

Methodology and approach

Data for this chapter was obtained through anthropological studies to explore how traditional belief systems contribute in natural resources conservation and management in Sankana and Tongo-Tengzuk. A case study approach was adopted, since it is appropriate for the study of the interaction between social actors and social phenomenon (Yin, 2003). The authors chose this study area because, aside from their contribution to natural resources conservation and management, very little has been done in terms of research. Data was drawn from primary and secondary sources. These include journal articles, books, District Assembly documents, internet publications, focus group discussions and personal interviews. The research was conducted from September 2014 to March 2015.

Conservation

Conservation and management are among the most important elements of sustainable development. It is the management of valuable natural resources such as timber, fish, topsoil, and minerals, forests, wildlife, parkland, and wilderness and watershed areas (Rim-Rukeh, Irerhievwie and Agbozu, 2013). There is no single definition of conservation; however, several definitions have been coined for the concept, with some scholars stressing the structural roots of anthropological interests, while others depart from the point of view of economic reasons. Peter Usher (2000) defined conservation as the maintenance of genetic species, and ecosystem diversity in the natural abundance in which they occur. W. H. Thomas (2003) also thinks conservation is the sacrificing of immediate rewards in return for delayed ones. Meanwhile, Eric Smith and Mark Wishnie (2000) see conservation as actions that prevent or mitigate biodiversity loss and designed for such purpose. We, therefore, define conservation as the maintenance of the ecosphere; water bodies, fisheries, habitats and biodiversity, etc., including protection and management. Conservation of natural resources is, thus, the wise use of the earth's resources by humanity.

Totems and taboos

The term *'totem'* comes from a North American Indian language, which refers to vegetables or animals that are revered by individuals, particularly a group of people, as holy or sacred. Totems are considered as an emblem consisting of an object such as an animal or plant that serves as the symbol of a family or clan. *Taboo*, on the other hand, is derived from the Polynesian term *tabu*, which means 'forbidden' (Adu-Gyamfi, 2011). Taboos are inhibitions or bans resulting from social custom or emotional aversion, which are declared as sacred and forbidden by people. Taboos represent unwritten social rules that regulate human behaviour. Totems and taboos are used by different groups of people for different reasons. Whatever the reason(s) for such constraints or regulations, totems and taboos at least locally play a major role in the conservation of natural resources, species and ecosystems (Johannes 1978, 1982, 1984 *a, b*, Chapman 1985, 1987, Gadgil 1987, Gadgil et al. 1993 cited in Johan and Carl, 1997). For this reason, Sigmund Freud (2004) and Robert Alun (2005) explain that totems are reverenced by people who believe themselves to be of one blood, descendants of a common ancestor, and are bound together by common obligations to each other by a common faith. Totem animals are used to maintain two useful prohibitions; against killing the totem animal (patricide) and against having sexual relations between members of the same totem or clan (incest). Our interest in this work is with the former rather than with the latter.

The traditional belief in the spiritual properties and uses of natural resources has effects on the protection and improvement of the natural environment in many rural communities in Ghana. This means that indigenous traditional belief systems have significant roles in natural resources conservation and management.

A totem can be an animal, a plant or any other natural object believed to be ancestrally related to an ethnic group, clan, or family as a tutelary spirit which they attach deep feelings to. Members of these ethnic groups, clans, or family do not eat, kill or trap such animals, birds, or fish. When a totem dies or is accidentally killed, members of the group it represents would show respect by, for example, mourning and burying it as in the case of a human being (Lumor, 2009). This is because they believe the totem to be ancestrally related to them as a tutelary spirit. Whether there is any truth in such beliefs is not the important thing for us in this work. What we rather care about is the end result of such beliefs and practices in the conservation of the environment. Totems have been used basically to preserve humanity in parts of Africa; in many ways, they have led to the conservation of other life forms and the environment generally.

Traditional belief systems

Since time immemorial, local people have developed a variety of resource management practices and approaches that continue to exist in tropical Africa, Asia, South America and other parts of the world (Appiah-Opoku, 2007). Our focus in this work will be on two Ghanaian communities. The contributions of indigenous and local belief systems toward a better understanding of natural resources and their sustainable use and management have been documented in the scientific and grey literature in many domains: biodiversity conservation and wildlife management, customary marine resource management, rural development and agroforestry, traditional medicine and health, impact assessment and natural disaster preparedness and response (IPBES, 2013). It is, therefore, evident that the role of traditional belief systems in the conservation of a large number of elements of local biodiversity regardless of their use value dates back to early times (Shastri et al., 2002).

One of such numerous approaches to natural resources conservation and management is the use of indigenous belief systems: taboos and totems. These traditional belief systems are capable of protecting biodiversity species in particular and the environment in general as long as the local communities have an interest in them. Traditional African societies also observe environmental ethics that help in regulating their interactions with the natural environment (Shastri et al., 2002). African Traditional Religion (ATR) and cultural practices in most parts of African communities are environmentally friendly and sustainable, contributing so much to natural resource sustainability and conservation (International Institute for Environment and Development (IIED), 1992). In Africa, the traditional belief systems ascribe supernatural powers to objects called *gods* and *goddesses*. A. Rim-Rukeh, Irerhievwie and Agbozu (2013) reiterate that African traditional religion and belief systems lie in the abode of the gods and goddesses located in rocks, streams, ponds, trees, land or anywhere within the community.

In Ghana, indigenous belief systems, including taboos, rituals and totems, have played significant roles in the conservation and management of natural resources. In his work *Traditional and Indigenous Methods of Conserving Biodiversity*, Yaa Ntiamoa-Baidu (1991) identified taboos and totems as the key indigenous methods for conserving biodiversity in Ghana. He further indicated that social taboos exist in invariably all cultures in the country and represent a class of informal institutions, where traditional, religiously governed norms or taboo systems define human behaviour. Totems and taboos have been important objects within human societies, serving as a commemorative object of ancestry and have promoted natural resources conservation and management over time, especially in Africa.

In many rural communities in Ghana, traditional belief systems remain the prime factor guiding people's conduct towards the exploitation of the natural resources (Rim-Rukeh, Irerhievwie and Agbozu, 2013). In a similar vein, they have promoted traditional belief systems in tourism development, generating revenue for community development especially, in areas where certain animals and other elements have been sacred. African traditional belief systems have latently been discovered to be one of the informal institutions geared towards the preservation of forest and other natural resources (Eneji et al., 2012).

However, in spite of the role(s) played by traditional belief systems in conservation of biodiversity and the environment, they are not given the needed recognition. Christianity and modernisation for instance, have worn away the significant role(s) played by these traditional beliefs in natural resources conservation, management and community development. Traditional religious belief systems are being eroded via acculturation and enculturation of most African communities, especially the introduction of Christianity (Eneji et al., 2012). In a similar vein, Seth Appiah-Opoku (2007) argues that modernisation sees ATR belief systems, worship and practices as rather inimical to the growth, unity, peace and cohesion of communities. Meanwhile, some of these changed beliefs and worship systems were caused by modernisation. These erroneous assumptions have retarded the growth and sustenance of the traditional African ways of environmental conservation.

The debate between Western and African belief systems on natural resource conservation and management

Since the pre-colonial era, indigenous knowledge systems (IKS) or ATR in natural resource management has been part of Africa's heritage, developed in order to address various survival challenges. IKS are locally oriented and have survived the test of time. However, European settlers who colonised the continent sought to destroy, denigrate or marginalise them and replace them with Western views and approaches, which were in line with their selfish goals of imperialism (Mapira and Mazambara, 2013). Before enslavement and the introduction of colonialism, the indigenous people of Africa were much concerned about their natural environment. They conserved and managed their natural settings for so many years. No one could mismanage his/her natural environment and go free without the anger of the gods. The anger and punishment that were arranged for persons who mismanaged their natural resources were severe; thus, community members were mindful in utilising the available resources. Therefore, certain areas such as woodlands, water points, mountains and certain animals, among others, were considered sacred and were not to be abused.

Above all, certain natural resources could be used only after rituals were performed by the fetishes. These practices helped to conserve the natural resources, and other living things, including fishes, wild animals, trees, forests and water bodies. Donnelly-Roark Paula (2004) observed that exploitation of the natural resources is systematic and done through the inspiration of spirit mediums and through the guidance of traditional institutions. Various indigenous beliefs and practices have contributed to indigenous natural resources management. Traditional institutions have played key roles in ensuring that those who broke natural resources management rules were punished.

These collective actions in natural resources management were expressed through religious beliefs and moral sanctions, and a range of sacred and cultural practices. The resilience of these beliefs and practices stood the test of time in natural resource management using taboos, totems, traditional protected areas, moral sanctions and rules and regulations (Millar, 2004).

On the contrary, Westernisation proponents would argue that the reason why Africans did not interfere with the natural environment for economic growth and development was that they were living in primitive societies and lacked innovative thinking, technology and the ability to advance. This claim can, however, be faulted in the sense that traditional strategies for environmental conservation like totems and taboos actually suggest deep thinking. If humans would not willingly respect the environment, then exploiting the advantage of what they respect so much in order to make them protect the environment seems like a prudent philosophical choice. Whether or not Africans lived in ignorance, at least they must be giving the needed applause for devising very effective means for conserving the environment for posterity. Colonisation and Westernisation through formal education and Christianity may be blamed for causing distractions to certain rich traditional systems relevant for natural resources management. Formal education and Christianity have somewhat eroded the rich cultural values and religious diversities of African people in Africa and have also changed their beliefs and worship systems (Smith and Wishnie, 2000). For instance, Appiah-Opoku (2007) reckons that prior to colonisation, indigenous African rulers occupied a unique position in the management of natural resources. They were accepted by their people as the custodians of religious, political, and judicial and the spiritual values of their communities, with the responsibility of managing both natural and community resources. However, the same cannot be said of today's African rulers and their population.

According to Ramadan Belhag and Yassin El-Kabir (1986, p. 251), early missionaries in Africa "condemned African customs and institutions and taught social norms of 19th-century Europe as though they crystallised a moral code of universal validity." Colonisation is hence a major cause of Africa's departure in their mode of natural resource management (Fairhead and Leach, 2004). James Fairhead and Melissa Leach (2004) reiterated further that decades of colonisation in Africa have alienated its people from their traditional ways of managing and utilising their natural resources. In Ghana, the missionaries separated their converts from their local communities to distant places called Salems or quarters. In the name of formal education, these converts were taught to frown upon their cultural beliefs and practices, including the observance of sacred days and adherence to the taboos (Adu-Gyamfi, 2011). The consequences have been that, through Christianity, Islam and civilisation in Ghana, these beliefs, taboos, customs and traditions have been relegated to the background and are regarded by many, especially Christians, as fetishes and useless, demonic and satanic, though they play a key role in environmental protection.

Taboos and totems in the management and conservation of natural resources at the community level

Studies have shown that in Ghana people protect their natural resources with taboos and totems (Abayie-Boaten, 1998). David Millar (2004) argues that the spiritual world in Ghana is the major driving force that regulates the performance of all traditional institutions in their quest to manage natural resources. For instance, the use of totems cut across all the cultures and they have historical significance as well.

Wildlife species regarded as totems have their historical or socio-cultural significance, as well as their symbolic quality of clans that adhere to it. The above discussion is supported by other studies (see Quarcoopome, 1987) by saying that animals are chosen as totems because they are said to have qualities which people would want to emulate; therefore, people identify themselves with such animals. Almost all the ethnic groups in Ghana have similar totemic animal symbols, but different meanings that they ascribe to them. However, this article limits its scope to the Sankana and Tongo-Tengzuk areas.

Sankana community

Sankana is located in the Upper West Region of Ghana. It is approximately 14 km from Wa, the regional capital, and it covers about 322 hectares of the total land area of the district (Issifu et al., 2013). Dagaabas and Sisaales are the two major tribes in Sankana. The Dagaabas constitutes 94% of the total population and the Sisaales represents 4% of the population. However, other ethnic groups such as Frafras, Walas and Akans constitute 2% of the total population (Issifu et al., 2013). The community has taboos and totems, which guide the behaviour and the conduct of the people. These taboos include but are not limited to whistling in the night and killing and eating frogs and pythons. In the case of whistling, it is believed to invite evil spirits. The focus of this chapter is on the forbidding of inhabitants from killing and eating pythons and frogs.

According to oral tradition, a python was believed to have turned into a log and helped the ancestors/forefathers of the community to cross a river during an attack by their enemies (TTFPP, 2011). During the enslavement era, raiders, including *Babatu* and *Samori*, most often attacked and captured the indigence of Sankana for the enslavement ordeal and other economic activities (TTFPP, 2011). The people used several strategies to prevent these raiders from attacking them but to no avail. Their population started diminishing because of the frequent attacks and the capturing of energetic men and women.

The people, through their traditional leaders, finally decided to migrate to a safer place to help shun the raiders. At a point in their journey, it became difficult for them to go any further, because there was the need to cross a river. In the periods of lamentation and wondering, they found a log on the river that helped them to cross over the river. After they had all crossed the river, it became known that the log that helped them to cross the river was not an ordinary log but a python. Since then the ancestors vowed not to kill or eat pythons, because they were considered saviours and helpers, hence, their totem. It is therefore believed that should anyone from the group kill or eat a python they will transform into a python and probably die, simply because of the link they have with them. This confirms Quarcoopome's (1987) earlier assertion that totems are the link between a person or group of persons and animal objects or a group of animal objects.

For this reason, there are many pythons found in and around caves in some sections of the community. These pythons are most often visited by tourists, and sometimes community members during festivals. These caves also served as places of refuge for their ancestors to hide from their enemies in the times of war. The caves again served as bunkers for the militants to fight the raiders and offered them protection against wild animals. As Kawano (1999) stated, "the people of Sankana would run into large caves to hide and mobilised ideas to fight off the raiders and seek an impenetrable shelter." Gunshots and hunting are prohibited at the cave sites because such acts are believed to invite the wrath of the cave gods, who prohibit noise making; hence such a prohibition in and around the cave sites has become a prospect for wildlife conservation.

A related story has it that a frog once led their ancestors to a source of water during difficult times in their migratory eras. It is said that their ancestors did not have enough water to drink in order to continue their journey, while wondering about how to survive in the coming days, a frog from nowhere jumped to where they were seated and jumped back to its initial base. The elders among the group decided to trace the origin of the frog and to their surprise found a water source. Henceforth, they have tabooed the killing and eating of frogs and have considered frogs to be their totem. In effect, water bodies where frogs are located are not misused. The proper utilisation of the water bodies not only prevents pollution, but also provides an enabling environment for fishes and other organisms to survive in good health, and it also serves domestic purposes as well, hence promoting natural resource management and conservation.

Tongo-Tengzuk community

Tongo-Tengzuk is located in the Talensi district of the Upper East region of Ghana. The Tongo Hills and Tengzuk Shrine are sited in Tengzuk village, 15 km southeast of Bolgatanga, the regional capital. It is bordered to the north by the Bolgatanga municipality, to the south by the West and East Mamprusi districts (both in the Northern Region), to the west by the Kassena-Nankana district, and to the east by the Bawku West and Nabdam districts (Ghana Statistical Service, 2014). The oral traditions of Tongo-Tengzuk claim that their ancestors have always been there, or, alternately, sprouted from the ground or descended from heaven over several decades ago (Gabrilopoulos, 1995).

The Tongo-Tengzuk communities hold the belief that crocodiles are human beings among them. It is therefore a taboo for any Talensi to kill or injure a sacred crocodile. Every Talensi, for that matter, knows that crocodiles are the incarnation of their important ancestors. To kill a crocodile is tantamount to killing a human being. As the story goes it is considered a murder, one of the most heinous kind and it would bring disaster to the clan if anyone from your family kills a sacred crocodile. For this reason, people are strictly careful as far as these sacred crocodiles in special pools are concerned. The presence of crocodiles in many water bodies makes people mindful to manage their water sources so that the crocodiles are not found wanting. Due to that, harmful chemicals are not used for fishing in and around the water bodies in the community.

In addition, rocks and hills are conserved and people (even the local folks) are prevented from using them for domestic and commercial uses. This stems from the belief that the rocks are the dwelling homes for most of their lesser gods. The conservation of the Tongo-Tengzuk cultural landscape site is therefore tied up with this worldview. The rock sites contain sacred shrines, including earth shrines (*tengban*) and ancestral shrines (*ba'ar*). The Talensi people worship the Supreme Being through the lesser gods, who are believed to be inside the rocks, cliffs and caves, and they constructed ancestral shrines. Paramount among the earth shrines is the *Tonna'ab* nestled in the cliffs to the west of the settlement at the section called *Kpatari*, which has provided the area with enormous sacred power for both the Talensis and other ethnic groups in Ghana. It has been viewed historically by the Akans in southern Ghana as a site of potential ritual power and, as such, Rattray (1932) observed the presence of wealthy Asante business persons in the area. *Tonna'ab* is a benevolent, protective and curative shrine or god that abhors evil. It has been the unifying factor for the Talensi and continues to flourish up till today.

At Kpatari, a suburb community, there are two sacred groves, *Bonab* and *Nnoo*, that greatly influence the culture of the Talensi. *Bonab* performs the same ritual functions as *Tonna'ab*. The *Nnoo* shrine serves as the home of the *Golib* god. The Golib god has a lot of influence on the agricultural affairs of the community by ensuring the maintenance of several cultural practices in the area. The Gologo festival is celebrated to reinforce the community's belief in the *Nnoo* shrine or *Golib* god. The historical establishment of and adherence to many taboos and norms in respect to the needs of the *Golib* god has helped to evolve and maintain a number of traditional conservation practices which, over the years, have preserved this unique cultural landscape. These taboos, norms and penalties have strengthened the relationship between the Talensi and their god or shrine.

The numerous trees in and around the shrines are not cut for firewood or charcoal (a major source of biofuel) but are left in their natural state because of the belief that lesser gods are found in and around the area. Also, the rocks of the hills are exempt from quarrying because of the presence of smaller gods in them. The presence of the lesser gods in the rocks found on the Tongo hills and the trees found in and around Tengzuk Shrine have all helped to preserve the natural environments from mismanagement. Hence, taboos and totems among the indigenous people of the Tongo-Tengzuk community have contributed immensely as far as natural resources conservation and management is concerned.

Lessons from taboos and totems

Wildlife conservation

Patches of forests in the communities are protected because they support sacred totems or animals believed to have special spiritual or cultural values and associations. For instance, the prohibition of killing and eating of pythons and frogs and crocodiles among the people of Sankana and Tongo-Tengzuk, respectively, has helped to conserve and maintain some forms of wildlife in the community, albeit not a well-structured wildlife zone. It has also helped the animals to survive and multiply in their numbers over time. Moreover, for the few tourists who go to observe some of the pythons and the caves located in some sections of the community, the location and historical backgrounds are also useful for academic purposes.

Sacred watercourses and biodiversity conservation

Tongo-Tengzuk is a masterpiece of a cultural landscape, one that the Talensi community has attempted over the years to preserve for posterity, despite increasing threats from modernity. Many rivers and streams that provide sources of drinking water for the Talensi specifically, and Tongo, are considered sacred because of the presence of crocodiles. The surrounding forestlands are protected on the basis that the lesser gods and other living spirits reside in the trees found in and around the Tengzuk shrines. As the crocodiles are intended to protect water bodies from being mismanaged, plants and animals, including fishes living in the water, are not harvested indiscriminately. No matter the reason for such practices, one certain thing is that biological diversity (flora and fauna) is conserved.

Conclusion

The focus of this chapter was to explore the significance of indigenous African belief systems: taboos and totems in natural resources conservation and management in Ghana with special reference to the Sankana and Tongo-Tengzuk in the Upper West and Upper East Regions of Ghana respectively. Taboos and totems have helped to promote not only the natural environment, but also traditional belief systems, as well as the generation of revenue for the communities through tourism activities. Indigenous African belief systems in the study area, including taboos and totems, have contributed immensely and effectively to the reduction of incidences of wildlife and biodiversity loss. Examples are the prohibition of killing and eating of frogs, crocodiles and pythons, which are totems for both the Sankana and Tongo-Tengzuk communities. More so, harmful chemicals such as dichlorodiphenyltrichloroethane (DDT), a synthetic insecticide which is highly toxic towards a wide variety of insects as a contact poison, are not used for fishing because of their potential danger to the lives of their totems living in the water; and because of that pollution of the water bodies, they are preserved in order to avoid the wrath of the gods believed to be living in them.

Cutting down of forest trees for charcoal and the quarrying of rocks in these areas have also been prevented because of the belief that lesser gods and other spirit beings reside in them. These, among other things, are believed to have helped in the promotion of natural resource conservation and management for a very long time. Despite the significant roles that indigenous/traditional African belief systems play in natural resources conservation and management, modernisation, Christianity and formal education have all become the greatest opponents of traditional belief systems. Therefore, religious leaders, including pastors, priests, imams, traditional authorities, etc., are encouraged to preach more about the sections of the Bible and the other belief systems that see human existence as closely tied to its natural environment, and thus, the provision of other environmental services which people need for survival on earth.

More so, the government, through the Ghana Tourism Authority, District Assemblies and the Ghana Museums and Monuments Board, should embrace a conscious effort to integrate modern laws and traditional norms and beliefs in natural resource conservation and management into a single policy framework enforceable by law. By doing so, rural communities within areas where these resources are located will willingly and voluntarily participate in the management and conservation of the natural resources for the benefit of the living and the yet unborn.

Finally, there is an increasing research interest in the benefits of African belief systems, especially in the conservation of the natural environment. This work is an attempt to further bring this underexplored part of the African knowledge system to the fore. We showed in this work the importance of a deeper understanding of how and under what conditions local communities adopt these systems in a response to environmental degradation, along with the root challenges involved. Hence, more research should be done to document the socio-economic and practical benefits of historic sites, traditional beliefs and practices in environment conservation.

References

Abayie-Boaten, A. (1998). Traditional Conservation Practices: Ghana's Example. *Institute of African Studies Research Review*, 14(1), 42–51.

Adu-Gyamfi, Y. (2011). Indigenous Beliefs and Practices in Ecosystem Conservation: Response of the Church. *Scriptura*, 107, 145–155.

Alun, J. R. (2005). *The Secret of the Totem*. New York: Columbia University Press.

Appiah-Opoku, S. (2007). Indigenous Beliefs and Environmental Stewardship: A Rural Ghana Experience. *Journal of Cultural Geography*, 22, 79–88.

Attuquayefio, D. K., and Fobil, J. N. (2005). An Overview of Wildlife Conservation in Ghana: Challenges and Prospects. *West African Journal of Applied Ecology*, 7, 1–18.

Attuquayefio, D. K., and Gyampoh, S. (2010). The Boabeng-Fiema Monkey Sanctuary, Ghana: A Case of Blending Traditional and Introduced Wildlife Conservation Systems. *West African Journal of Applied Ecology*, 17, 1–10.

Belhag, R. S., and El-Kabir Yassin, A. (Eds.) (1986). *Christian Missionarism and the Alienation of the African Mind*. Tripoli: Dar Iqra.

Eneji, C. V. O., Ntamu, G. U., Unwanade, C. C., Godwin, A. B., Bassey, J. E., Willaims, J. J., and Ignatius, J. (2012). Traditional African Religion in Natural Resources Conservation and Management in Cross River State, Nigeria. *Environment and Natural Resources Research*, 2(4), 45–53.

Fairhead, J., and Leach, M. (2004). False Forest History, Complicit Social Analysis: Rethinking Some West African Environmental Narratives. Samantha Jones and Grace Carswell (Eds.), *Environment, Development and Rural Livelihoods*. London: Earthscan, 12–31.

Fabricius, C. (2004). The fundamentals of community-based natural resource management. Fabricius, C., Koch, E., Magome, H., and Turner, S. (Eds.), *Rights, resources and rural development: Community-based natural resource management in Southern Africa*. London: Earthscan, 3–43.

Freud, S. (2004). *Totem and Taboo*. New York: George Routledge and Sons.

Gabrilopoulos, N. (1995). *Ethnoarchaeology of the Tallensi Compound* (Upper East Region, Ghana). A Master of Arts thesis submitted to the Faculty of Graduate Studies, University of Calgary, Canada.

Ghana Statistical Service (2014). 2010 population and housing census. Talensi District Analytical Report (online). Available at: http://www.statsghana.gov.gh/docfiles/2010_District_Report/Upper%20East/TALENSI.pdf. DA: 21/02/2015

Hulme, D., and Murphree, M. (1999). Communities, Wildlife and the 'New Conservation' in Africa. *Journal of International Development*, 11, 11–285.

International Institute for Environment and Development. (1992). *Environmental Synopsis of Ghana*. London: IIED.

Issifu, A. K., Antwi, S., Kwofie, I., Quarshie, G., & Mustapha, H. (2013). *The role of parents in promoting girl child education in Sankana*. Unpublished BA degree thesis submitted to the Department of Community Development, Faculty of Planning and Land Management. University for Development Studies, Wa Ghana.

Johan, C., and Carl, F. (1997). The Relations Among Threatened Species: Their Protection, and Taboos. *Conservation Ecology* [online], 1(1), 6. www.consecol.org/vol1/iss1/art6/

Kawano, M. (1999). Product of the Past: An Examination of Ashanti and Dagari Proverbs as a Means of Exploring the Cultural and Societal Effects of the Trans-Atlantic Slave Trade on the Dagaaba People and the Ashanti People. *African Diaspora ISPs. Paper 80*.

Lumor, F. (2009). *Significance of Animal Symbolism Among the Akans of Akyem Abuakwa Traditional Area*. Unpublished Master of Arts in Art Education submitted to Faculty of Fine Art, College of Art and Social Sciences, KNUST-Ghana.

Mapira, J., and Mazambara, P. (2013). Indigenous Knowledge Systems and Their Implications for Sustainable Development in Zimbabwe. *Journal of Sustainable Development in Africa*, 15(5), 90–106.

Millar, D. (2004). *Traditional African Worldviews From a Cosmovision Perspective*. Compas Series on Worldviews and Sciences 3. Tamale: COMPAS/UDS/UCC.

Ntiamoa-Baidu, Y. (1991). Conservation of Coastal Lagoons in Ghana: The Traditional Approach. *Landscape and Urban Planning*, 20, 41–46.

Ntiamoa-Baidu, Y. (1995). Indigenous vs. Introduced Biodiversity Conservation Strategies: The Case of Protected Area Systems in Ghana. *African Biodiversity Series No. 1*, Biodiversity Support Program, Washington, DC.

Ostrom, E. (1990). *Governing the Commons: The Evolution of Institutions for Collective Action*. Cambridge: Cambridge University Press.

Overseas IPBES (Intergovernmental Platform on Biodiversity and Ecosystem Services). (2013). *Initial Elements of an IPBES Approach: Towards Principles and Procedures for Working With Indigenous and Local Knowledge (ILK) Systems*. IPBES/2/INF/1/Add.1

Paula, D. (2004). *Indigenous Knowledge Systems in Sub-Saharan Africa: An Overview*. Rome: Food and Agricultural Organisation.

Quarcoopome, T. N. (1987). *West African Traditional Religion*. Ibadan: University Press.

Rattray, R. S. (1932). *The Tribes of the Ashanti Hinterland* (Vol. II). Oxford: Oxford University Press.

Rim-Rukeh, A., Irerhievwie, G., and Agbozu, I. E. (2013). Traditional Beliefs and Conservation of Natural Resources: Evidences From Selected Communities in Delta State, Nigeria. *International Journal of Biodiversity and Conservation*. www.academicjournals.org/IJBC

Roe, D., Nelson, F., and Sandbrook, C. (Eds.) (2009). Community Management of Natural Resources in Africa: Impacts, Experiences and Future Directions. *Natural Resource Issues No. 18*, International Institute for Environment and Development, London.

Shastri, C. M., Bhat, D. M., Nagaraja, B. C., Murali, K. S., and Ravindranath, N. H. (2002). Tree Species Diversity in a Village Ecosystem in Uttara Kannada District in Western Ghats, Karnataka. *Current Science*, 82, 1080–1084.

Smith, E. A., and Wishnie, M. (2000). Conservation and Subsistence in Small-Scale Societies. *Annual Reviews Anthropology*, 29, 493–524. http://dx.doi.org/10.1146/annurev.anthro.29.1.493

Third Trimester Field Practical Programme TTFPP. (2011). *Sankana Community Profile by Group 292. FPLM/DCD*. Ghana: University for Development Studies, Wa Campus.

Thomas, W. H. (2003). One Last Chance: Tapping Indigenous Knowledge to Produce Sustainable Conservation Policies. *Futures*, 35, 989–998.

Yin, R. K. (2003). *Case Study Research: Design and Methods*. Applied Social Research Methods Series (Vol. 5, 3rd ed.). London: Sage Publications.

Index